WALES THE MISSING YEARS

TALES FROM AN AREA THE SIZE OF WALES - 4,567 TO 700 MILLION YEARS AGO

JOHN S MASON

Word after Word

Published by Word After Word Press, 2021
FE20211

Word After Word Press
www.wordafterwordpress.com

ISBN 978-1-9161655-1-9

A CIP catalogue record for this book is available from the British Library.

Design and layout by Sara Holloway Graphic Design - www.saraholloway.com
Printed and bound by Instantprint

Special thanks to Arts Council of Wales

In memory of

Shena Mason

(1938-2014)

CONTENTS

INTRODUCTION 11

DEEP TIME 14

WALES: THE MISSING YEARS

PART ONE: BORN OF EXPLODING STARS

16 A late December night on Borth Beach
20 Star birth
21 Matter under the microscope
25 Stars and their life cycles: heavy metal factories
28 Big stars and big bangs
31 Density beyond comprehension
34 Meanwhile back on the beach

PART TWO: ORIGINS

36 From interstellar dust to Planet Earth
38 Science in a nutshell
40 A space rock from the deep past
44 Planet birth
47 Meltdown!

PART THREE: EARTH AND MOON - THE HADEAN EON
53 4,560 to 4,000 million years ago
60 Seas of Magma
65 Architecture of a planet: Earth's vital magnetic field

PART FOUR: INTRODUCING THE PRECAMBRIAN - TERRAFORMING
70 4,000 to 541 million years ago
75 Metamorphism
81 Water - an alien chemical?

PART FIVE: EARTH'S GREENHOUSE EFFECT
88 A paradox solved

PART SIX: LIFE ON EARTH
96 When and how?
101 Recipe for amino acid soup

PART SEVEN: PLATE TECTONICS
104 The facilitator of biodiversity
107 Magnetic attraction
112 Subduction - how oceanic crust meets its fate
117 Making mountains

PART EIGHT: THE GREAT OXYGENATION EVENT, THE OZONE LAYER AND EARTH'S CARBON CYCLE

120 What could possibly go wrong?
125 Earth's vitally important carbon cycles, fast and slow
130 Why the sea is salty
132 Fossilised calcifying organisms
133 Ocean acidification explained
136 The Siberian Traps
142 Snowball Earth - what happened?

PART NINE

148 The missing years come to an end

EPILOGUE 150

INTRODUCTION: A small but well-travelled country

"The world's three great units of measurement are the mile, the kilometre, and the size of Wales."

Thus began an article on the popular Welsh media website, Wales Online, posted on 19th September 2003. And it's true: the unit has been used in all sorts of guises. To cite but three examples: the Laikipia wildlife reserve in Kenya has been proudly described as being the size of Wales. More negatively, Sky News reported from the Amazon Basin in 2013 that, "the worst year on record was 2004, when 27,000 sq kms (16,777 sq miles) of forest was destroyed; an area roughly equivalent to the size of Wales." On a happier note, there is an environmental charity called Size of Wales which has the goal of protecting a total area of rainforest of that size. Good for them. But what of the place that inspired the creation of this popular unit of measurement? What is its story?

With a surface area a little under 21,000 square kilometres, Wales is a small country on the Atlantic fringe of north-western Europe. What it lacks in size, however, it more than makes up for in its scenery, its culture, the richness of its language and its heritage. The heritage of Wales includes both its history and its prehistory.

History, according to my battered old copy of the Oxford English Dictionary, is the "methodical record of public events; past events, course of human affairs". The implication is that history is mostly the things we witnessed and recorded by writing about them. On that basis, everything that happened before we started writing about stuff can be lumped together under the term, 'prehistory', which is the precise subject of this book.

This volume covers, so far as Wales is concerned, its Dark Ages: the span of geological time between the formation of the Solar System and the oldest known Welsh rocks. That span began 4.57 billion years ago and did not come to an end until around 700 million years ago. For Wales, these are the missing years. A lot of missing years, in fact. So why are these ancient and far off times part of its story?

It's because the chemical elements and compounds, making up the rocks, rivers, seas and skies of Wales, plus its plants, animals and people, have a far longer history than the country itself. Such materials are present for one reason alone: a consequence of everything that happened in the past, up to and including the formation of Planet Earth. This common origin is shared by everywhere - and indeed everyone - on Earth. Their tale is most certainly part of Wales' story, as are the facts that we are blessed with breathable air, drinkable water, skies free of deadly radiation (if one uses sun cream) and a survivable climate (if we don't go and wreck that). All of these are things we share with the biodiversity around us. However, these are also things we tend to take too much for granted. I hope that by reading this book, you will come to realise that the fact we are here at all is down to repeated episodes of extraordinarily good fortune, along the geological timeline. It is likewise hoped that such realisations will lead to a renewed appreciation of Planet Earth – our only home.

Wales: The Missing Years, is my story, but it belongs to every single one of you, too.

- John S Mason, Machynlleth, Wales, 2021.

DEEP TIME

Geological time is difficult to imagine. When the timeline is more than 4,560,000,000 years long, it's hard to visualise: as with the light-year, the unit of measurement of distance in outer space, the numbers involved are too far outside of our everyday perspective. We need context, something that can link deep time to a more familiar scale. How about if we take all of geological time from the formation of Planet Earth through to the present day and plot it against the twelve month human calendar? That works surprisingly well. On the resulting chart, Earth is formed on January 1st as Big Ben starts to chime. The present day is the point at which the New Year's Eve revellers are ending their countdown. Everything in between, from winter through spring, summer, autumn and into winter again, represents the geological timeline.

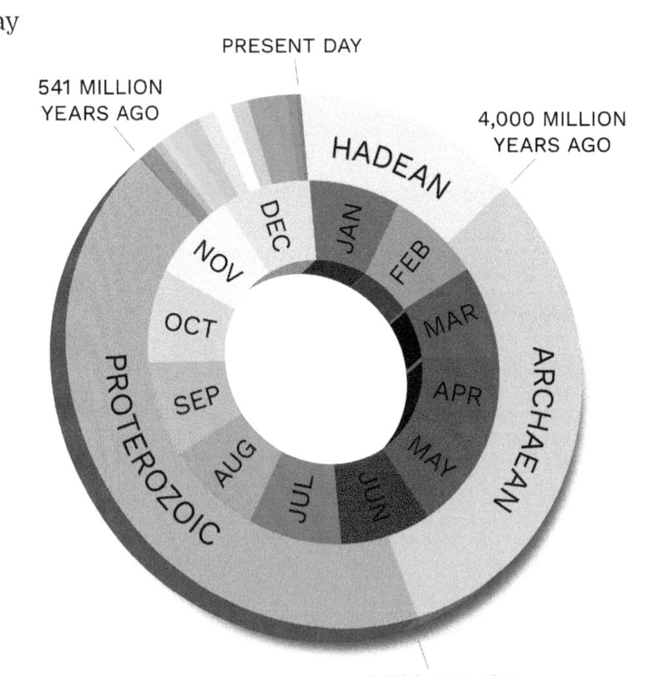

To make life easier for geologists, geological time has divisions and subdivisions. 'Eons' are the big divisions, representing half a billion years or more. Eons are subdivided into 'Eras', each lasting a few hundred million years. Eras are split up into 'Periods', typically a few tens of millions of years long and Periods are in turn divided into 'Epochs'. Epochs themselves consist of two or more 'Stages'.

Each stage has, or will eventually have, an internationally agreed 'type section', a good quality outcrop of rocks somewhere on the planet, where the boundary between that stage and the one preceding it is clearly defined. I say, "will eventually have", because in some cases geologists are still arguing about them, but many type sections are now firmly agreed, especially with respect to the Stages of the last few hundred million years.

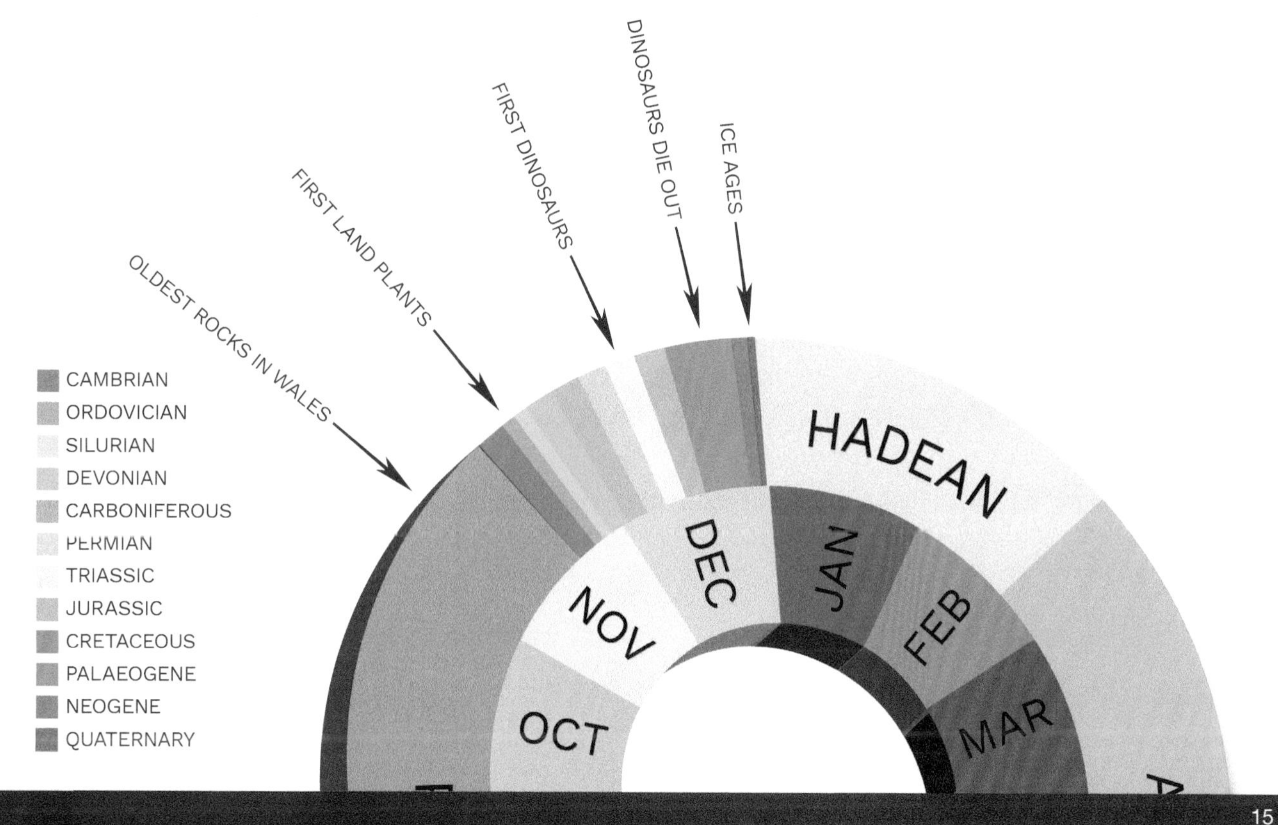

WALES THE MISSING YEARS

Part one: Born of exploding stars

A late December night on Borth Beach

Darkness gathers fast over the lonely winter shore. One by one the lights of the coastal towns of Cardigan Bay come into view: Aberystwyth, Tywyn, Aberaeron, Ceinewydd, Criccieth and then Pwllheli, isolated oases of brightness twinkling in a black desert. White surf-crests fade and merge into the twilight: all that remains is sound. Somewhere out there, the call of an oystercatcher echoes across an unknown distance.

It's good to be out fishing on a night like this, when the weather gods are, for once, playing nicely. Instead of cowering from rain or hailstones, I can stand contentedly, watching the tips of my rods in their rest, both clearly visible in the light of my head torch, thanks to their coatings of reflective paint. Carefully tensioned monofilament nylon lines, a fraction of a millimetre thick, connect each rod and its high capacity reel to a set of end-tackle, or rig, cast out to where we think the fish may be feeding, maybe 50 or 100 metres or more out from the shore. Rigs vary a lot; there are all sorts of designs, but their main purpose is to present the baited hooks on or above the seabed, where, we hope, the fish will find them.

Once out there, the streamlined 140 gram lead fishing-weight, that towed the rig up through the air during the cast, has to take on a new role. It has to hold the rig in place on the seabed. If the tide is strong, holding may require using specially designed weights equipped with stiff steel wires that will dig into the sand like an anchor. Fishing, in its modern form, does rely on a wide range of technologies and materials, all working in support of one another. As a consequence of all that technology working hand-in-hand, so long as I concentrate from now on, I will immediately notice if I get a bite. The fish will pull on the tensioned line which, in turn, will cause the relevant rod-tip to move.

Such movements vary in character, according to the species of fish making off with the bait. Small flatfish, like dabs, produce rather half-hearted rattles. At the other end of the fishy spectrum, a big bass or a ray can yank the rod-tip hard down or even heave the rod right out of its rest, so it's a good idea to pay attention. Concentrating means looking up a lot and that's no bad thing, under the clear, light pollution-free skies of western Wales, where it can sometimes seem as though you are gazing out towards infinity.

Some minutes pass. A shooting star races fleetingly across the heavens. Nothing so unusual about that, but on occasion, we night fishermen witness more spectacular objects. I think of the brilliant fireball I saw a few years ago, passing high overhead. It split asunder, sparks flying, the two incandescent fragments then diverging as they blazed over the northern horizon. People fishing the Bristol Channel coast saw it as well, as did those chasing the cod further north in Cumbria, something confirmed next day by a trawl through online sea angling forums. There's always plenty to see, out there, away from the nagging insistences of display screens and the myriad other distractions of modern life.

Gazing out towards infinity: star systems span the caverns of the night, across inconceivable distances. Out there, all around me, a never-ending stream of extraordinary events is taking place: the Universe never rests. Take, for example, the constellation of Orion, the Hunter of Greek mythology, coming into view by mid-evening, as it rises high above the southern horizon. Orion's Belt – consisting of the three stars Alnitak, Alnilam and Mintaka, all in a straight line – is one of the most familiar objects in the night sky.

Orion - The Hunter →

Below Orion's Belt there hangs the Hunter's Sword – another, fainter line of stars. Within this small (to me) area of the night sky, there is also a little, fuzzy-looking patch of light. This is the Orion Nebula, or M42 as it is called by astronomers. It may look tiny, but that's because it is more than 1,300 light years from Earth. M42 is in fact some 24 light years, or 227 trillion kilometres, in diameter.

Like any other nebula, M42 consists of vast clouds of gases, dominated by hydrogen, along with interstellar dust and charged particles. Seen through a high-power telescope, it clearly has multiple brighter and darker regions, like an aerial view of a city at night. And it's a busy place for sure: it has been described by astronomers as a star factory. Within its swirling veils, new stars are continually being assembled, as colossal clumps of matter are brought together by primordial forces such as electromagnetism and gravity. These are everyday forces that we constantly experience here on Earth, but in this case acting on a vast, unimaginable scale.

Let's just take one such force – gravity. Gravity involves big things attracting little things. Just as one example, the gravitational pull of Earth is what makes you fall off a rock climb if you lose your grip. You are being pulled, by gravity, down towards the planet's centre, since Earth's core is an immensely bigger thing than you. That's a simple fact of physics that explains why many climbers prefer to use ropes and running belays, since they know what will happen if they let go, deliberately or accidentally. Such forces are present, not just here on Earth, but all around the Universe.

The Orion Nebula, taken using the HAWK-I infrared camera on ESO's very large telescope in Chile. →
Photo: ESO/H. Drass, et al.

Star birth

In any such star factory, as more and more matter is drawn together, the gravitational pull of an embryo star (or 'protostar') grows stronger and stronger. Eventually, it will attract everything in its near vicinity, drawing the matter in towards itself from all directions. As the matter becomes more and more densely packed together, it becomes hotter and hotter. Although it often feels unpleasantly hot on an overcrowded Tube train, that's nothing compared to star birth. Within the core of a protostar, as that matter continues to accumulate, the temperature climbs and climbs until it crosses a critical threshold. Let's give that threshold some context in terms of everyday things. Water freezes at zero degrees Celsius (°C) and boils at 100°C. Tin melts at 232°C, brass at 927°C and iron at 1,538°C. At the point of star birth, a typical temperature is around ten million degrees Celsius.

Once such a temperature threshold has been crossed, the protostar becomes an incandescent source of heat and light: in other words, a star, a new source of sunshine. For a star of a similar size to our Sun, the whole process, from growing dust cloud through to stellar adulthood, takes approximately fifty million years. Since the Universe is full of such nebulae, the process of star formation is going on all over the place, all the time. More than four and a half billion years ago, a very similar set of circumstances and processes came together right here, in our own part of the galaxy. The product was a Solar System, complete with a central star and all of the surrounding planets, comets, asteroids and so on.

We all know that sunshine is bright and warm, but why? In order to appreciate what stars actually do, we first need to think a bit more about matter, because stars are all about matter. Matter, whether in the swirling clouds of M42 or here on Earth, is another word for stuff. Everything around us, everything you can see inside and outside of your house, consists of matter. You, likewise, consist of matter, as do I and everyone else. So what are we made of?

Matter under the microscope

At one level of magnification, matter is made up from the chemical elements, of which almost 90 occur naturally on Earth. Elements can either occur alone, like the oxygen molecule (O_2), that essential-to-life component of the air that we breathe, or in various combinations, known as chemical compounds such as the water (H_2O) that we drink. The soils and rocks beneath our feet are made up of lots of different chemical compounds known as minerals, of which, at the time of writing, there are 5,566 different known varieties. Some minerals can be found almost everywhere, such as quartz, which is silicon dioxide (SiO_2). Milk-white pebbles of quartz are common along most UK beaches. At the other end of the scale, some of the rarest minerals on Earth may be known from just one or two samples, discovered at a single locality. Elements, too, vary a lot in their abundance on Earth. As we all know, there's a heck of a lot more iron around than there is gold, for reasons that we'll come to shortly.

Quartz can be seen as millk-white pebbles on many beaches.

Bright green Redgillite, forming millimetre-sized crystals, was only described as a new mineral in 2005. Photo: D.I. Green

Periodic Table of Elements

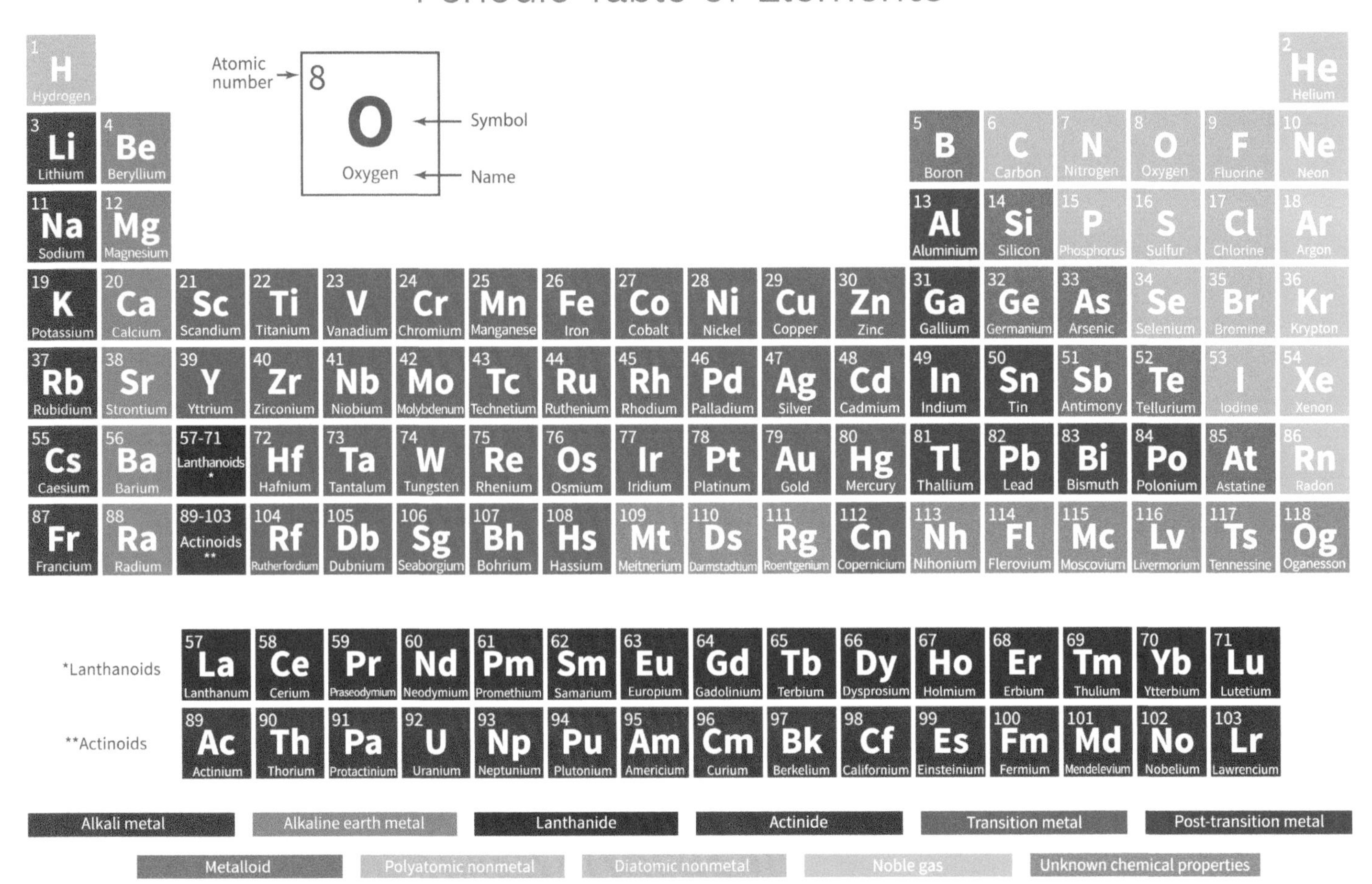

Zooming in further with our virtual microscope, any sample of any individual element, be it gold, iron or whatever, is made up of vast numbers of atoms. Atoms are tiny: we're talking about diameters typically around one millionth that of a human hair. Each and every atom consists of a central nucleus, surrounded by electrons. In turn, each and every atomic nucleus consists of subatomic particles, the most familiar of which are protons and neutrons. Protons each carry a positive electrical charge, neutrons have no charge whilst the electrons surrounding an atomic nucleus each carry a negative charge. In any element, the number of protons and electrons in each atom is the same, so that the electrical charges balance each other out. It's often said about human relationships that opposites attract, but in physics it's always the case.

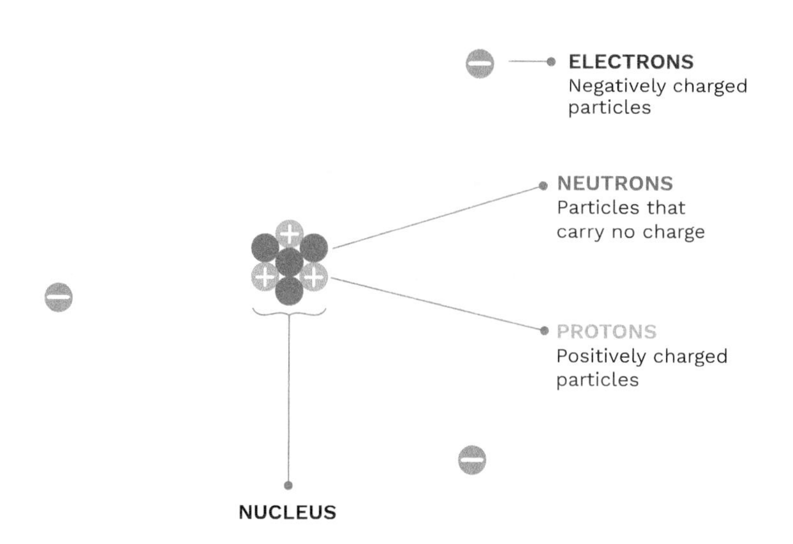

Each of the individual elements has a fixed atomic number which corresponds exactly to the number of protons in the nuclei of its atoms. Hydrogen is element 1. It has the smallest atom of all the elements, with just a single proton in its nucleus. Helium, element 2, has two protons in its nucleus. Atoms of carbon, element 6, have six protons in their nuclei, and so on up to uranium, the heaviest element occurring naturally on Earth, with 92 protons and therefore an atomic number of 92.

← A lithium 7 atom

While the number of protons is a fixed figure for the atomic nucleus of any given element, the number of neutrons can vary a little. As a consequence of this variation, most elements occur in two or more versions, each of which has a different number of neutrons from the others. In the case of carbon, there are three versions present in nature. Each version has six protons, but carbon 12 has six neutrons, carbon 13 has seven and carbon 14 has eight. The numbers 12, 13 and 14 are the total number of protons and neutrons in the nuclei of each version and are known as the "mass numbers". Different versions of any element, with their differing mass numbers, are known as 'isotopes'.

In nature, each different isotope of a given element can behave a little differently from the others. Incorporated into a chemical compound, they may make that compound dissolve more easily in water, or cause it to have a slightly higher or lower boiling point, to give just two of many examples. Some isotopes are mildly to fiercely radioactive: others are completely stable. Such features are extremely useful to Earth scientists. With modern analytical technology, we can determine the relative abundances of the different isotopes of specific elements (including carbon), within the rock samples we collect in the course of our research.

Isotopic compositions of rocks can tell us all sorts of stuff. They can give us valuable information about the environmental conditions in which a rock, such as a bed of limestone, was formed. Examples include things like the chemistry and temperature of ancient sea water or the composition of Earth's atmosphere at the time. Certain isotopes, such as those of uranium and lead, make it possible to determine the precise age of a rock. With tools like these at our disposal, along with the many others now available to geologists, the past can be reconstructed in ever-expanding detail. It's the next best thing we have to a time machine.

Stars and their life cycles: heavy metal factories

Of all the chemical elements present in the universe, hydrogen is by far the most abundant, accounting for almost three quarters of all atomic matter. Helium makes up almost a quarter, while the rest of the elements, from lithium (element 3) up to uranium, make up the tiny remainder. Hydrogen is pure star fuel. In stars, countless nuclei of hydrogen atoms are constantly being fused together to form helium atoms. The nuclear fusion reaction produces energy and light, better known as sunshine. It's the same kind of reaction that goes on when a hydrogen bomb is detonated, but in the case of stars it is a steady, self-controlled process. All the same, it's a difficult environment to imagine. In the case of our own Sun, the temperature at its centre is about fifteen million degrees Celsius.

Stars come in a wide range of sizes. The Sun is a rather average-sized star, with a diameter of a little over a million kilometres. Dwarf stars can be a fraction of that size, while giant stars can be hundreds of times bigger. Size matters in this case, since there is an astronomical rule of thumb that the larger the star, the shorter its lifespan. The smallest dwarf stars can last for many billions of years. In stark contrast, the largest stars have a life expectancy of just a few million years.

The fact that giant stars can be so short-lived is a critical factor in this story. Why? Consider for a moment such short life expectancies, set against the age of our home galaxy, the Milky Way, a respectable 13.6 billion years old. In all of that long, long time, countless larger stars must have already been born, shone through their lifetimes and died. To Planet Earth and its inhabitants, those previous stellar life journeys have been of critical importance. We would not be here without them, not one of us. That's a weird concept upon first acquaintance, so to figure out how, let's take a closer look at the way in which stars evolve and, eventually, die.

Any star has a 'main sequence', a scientific term referring to the time during which it steadily fuses its hydrogen fuel into helium. For a star the size of the Sun, the main sequence typically lasts for around ten billion years. Our Sun has now been in its main sequence for about 4.57 billion years, so it has plenty of time left. But to discover its eventual fate, let us return our gaze back to the night sky and the constellation of Orion. Forming the Hunter's left shoulder is the conspicuously bright red supergiant star, Betelgeuse. Geologically young at only ten million years old, Betelgeuse is a classic example of one of those bigger stars whose main sequence doesn't last very long. In fact, its main sequence ended some time ago.

Once the main sequence of Betelgeuse came to an end, meaning that it had used up most of its supply of hydrogen, the star underwent a radical transition, switching from hydrogen to helium fusion – 'reburning' the product of fused hydrogen. A star that is fusing together helium atoms is able to create still heavier elements containing even more protons and neutrons: carbon, oxygen (element 8) and so on.

During the transition from its main sequence to helium fusion, Betelgeuse underwent a massive expansion: if it occupied the centre of our Solar System, its outer edge would be somewhere near Jupiter. When one considers that the average distance between Jupiter and the Sun is 778.5 million kilometres, one can certainly appreciate the term, supergiant. The red appearance of Betelgeuse and other stars at this stage of their lives is because with their enormous surface area radiating energy, their outer layers are cooler – they are now only red-hot, as opposed to white-hot.

At the end of its main sequence, billions of years in the future, our Sun will likewise evolve into a red giant, but we have to remember that it has always been a much smaller star than Betelgeuse. Due to that size difference, there are major implications with regard to what happens next. In stars of up to 1.4 times the Sun's size, the completion of the helium-fusion stage will be the end of the line. The stupendous temperatures required to take the process further - to start fusing carbon atoms together - will never be reached. Our sun will run out of fuel: it will simply fizzle out. Its outer layers will drift away into space, leaving behind a small, planet-sized core. This forlorn remnant will become another example of what is known as a 'white dwarf', consisting of the densely packed products of helium fusion, hot at first but slowly cooling over many billions of years.

Big stars and big bangs

That's the ultimate fate of our average-sized Sun, but what happens to monsters like Betelgeuse? In much bigger stars, things get a lot more interesting. The deep interiors of such stars are much hotter and as the core temperature rises to mind-boggling levels, new nuclear reactions begin, involving those products of helium fusion such as carbon.

Carbon fusion begins at a temperature of around five hundred million degrees Celsius. The process is a bit more complicated than hydrogen fusion and produces oxygen, neon (element 10), sodium (element 11) and magnesium (element 12). Fusion of oxygen begins at getting on for two billion degrees Celsius and goes on to produce silicon (element 14). Eventually, it can get so hot that silicon itself undergoes fusion. Silicon fusion leads to the production of iron and nickel (elements 26 and 28 respectively). Importantly, iron and nickel are the heaviest elements that result from the regular nuclear fusion process, even in the biggest stars. Too much energy is required to take the fusion reactions any further.

Stable elements heavier than iron can however be produced by processes other than fusion, inside the larger red giant stars. One such process is known as 'slow neutron capture', initiated when free neutrons collide and merge with atomic nuclei. The greater the amount of free neutrons whizzing around, the more effective the process. Slow neutron capture involves some fairly complicated particle physics, but for the purposes of this story we only need to know one thing: that the end products are heavier elements. The process has its limitations, though. Bismuth (element 83) is the heaviest element that can be formed in such a manner. For making the full spectrum of heavy elements, we need to crank things up a notch. Events now start to get seriously violent.

The heavy end products of these complex nuclear processes naturally tend, through gravity, to gather towards the core of the star, now approaching its grand finale. Up until now, the immense heat and radiation, pushing outwards from the nuclear reactions in the core, had acted as a counterweight to that gravitational pull, thereby holding the star together. But at the point where the fuel supply of a giant star finally runs out, things change very quickly indeed. Core temperature falls away sharply and the great outward pressure, exerted by heat and radiation, suddenly weakens. Without that counterweight in place, gravity wins and the star passes a point of no return, the core collapsing in on itself, in a matter of seconds. Let's put some numbers on that: a star-core around 8,000 kilometres in diameter collapses to form an object maybe 20 kilometres across, in mere moments. The collapse causes a sudden temperature spike of up 100 billion degrees Celsius. Meet the supernova. So explosive are the forces released in such events that all the outer layers of the star are blown out into deep space in a colossal thermonuclear shock wave.

A supernova releases so much energy so rapidly that, for a brief time that may be no more than days or weeks, it can outshine an entire galaxy, like a suddenly ignited beacon in the dark depths of space. With that much energy around, other nuclear processes can thrive, such as 'rapid neutron capture'. Rapid neutron capture is vastly more efficient when it comes to generating heavy elements, compared to its slow counterpart. The atoms of such formed heavy elements will be flung out into space, like countless seeds, along with everything else making up the residual particle cloud. And left behind, in place of the star's core, there will be a tiny but phenomenally dense body of super-compressed matter. The precise nature of this remnant body will depend on the size of the original star. In the case of stars originally up to around three times bigger than our Sun, the collapsed core forms a neutron star. However, exploding stars of any greater size - and some are far bigger - leave behind black holes, bodies consisting of matter packed so densely that their gravitational fields are capable of pulling in anything in their vicinity - including light itself.

Density beyond comprehension

Neutron stars, too, severely challenge all of our everyday concepts of matter, based as those are upon the stuff we encounter here on Earth. They contain matter so densely packed that just a teaspoonful of it would weigh 900 times more than the Great Pyramid. You're not alone: I cannot for the life of me imagine that. It's too far beyond our normal frame of reference. Such density means that a neutron star, whilst not able to draw light towards itself like a black hole, nevertheless has a tremendous gravitational pull. At the surface of a typical neutron star, the force of gravity is thought to be some two hundred billion times stronger than what we are used to here on Earth.

Some neutron stars occur in pairs as binary systems, orbiting around one another, the remnants of a double bill of supernovae. This arrangement can lead to what must be the ultimate in cosmic fireworks. Binary neutron stars may spiral inwards towards one another through gravity, until eventually they merge together. The collision leads to one neutron star ripping the other apart and absorbing most of its matter; such marriages are another way in which black holes may be formed. Phenomenal amounts of energy are released in such a process, in the form of a brief but intense flash of radiation, known as a 'gamma-ray burst'.

Gamma-ray bursts are not visible to the naked eye because we cannot see that type of radiation. However, they can be spotted in distant galaxies, using instruments aboard satellites orbiting Earth. The satellite era is only a few decades old, yet we've already detected a fair few of them. It therefore follows that gamma-ray bursts pop up here and there around the Universe on an occasional but not infrequent basis. There must have been a lot of such events in total, going back and back through deep time.

← A mosaic image of the Crab Nebula, a six light year wide remnant of a star's supernova explosion, taken by NASA's Hubble Space Telescope.

Photo: NASA, ESA, J.Hester and A.Loll (Arizona State University)

As technology progresses, our observational instruments are able to peer deeper and deeper into space and in doing so, they have helped us make an important discovery. When neutron stars collide, the rapid neutron capture process, producing heavy elements, operates on a vast scale – far more effectively than during a supernova. Although supernovae do produce heavy elements such as platinum (element 78), gold (element 79), lead (element 82) and uranium, past neutron star collisions may have have generated the lion's share of such substances. Given that extraordinary concept, it's a bit ironic that today, some of us physically fight, or threaten to fight, over the first two elements using the latter two.

If we have to rely on occasional events like neutron star collisions to produce the bulk of the heaviest elements in the Universe, then it's not so surprising that such elements tend to be relatively uncommon. Large supergiant stars, fusing silicon into iron in the final stages of their life-cycles are, in contrast, commonplace objects. No wonder, then, that iron is vastly more abundant than gold, on Earth, in the Solar System and, indeed, throughout the cosmos. But how did our own Solar System come to incorporate such heavy elements when a star the size of the Sun cannot make them? Again, the answer has come from peering out into deep space and examining those nebular clouds.

The product of a dying star? Indeed. Gold forming millimetre-sized flakes in quartz, from North Wales.

Our observations of other nebulae have shown that stars tend to form in clusters. Most clusters will include giant stars that last only a few million years, before exploding and seeding their surroundings with countless atoms of the heavy elements. Likewise, many such clusters will, at some point, contain binary neutron star systems, primed for collision. Although our Sun is now a lone star, drifting along through its own corner of the Milky Way galaxy, there is little doubt that it originally came into being within such a cluster, in which there had already been plenty of supernovae, quite possibly with the odd neutron star collision thrown in, too. In other words, it was born into a noisy neighbourhood.

Just how the Sun ended up on its own is a scientific problem yet to be answered. Either the parent star cluster dispersed, or somehow the Sun was ejected from it. We'll leave that one for the astrophysicists to worry about. Either way, here we are, a long way from neighbouring stars, something that has been fortuitous, since the further we are from any giant stars, the less harm they can do to us when they explode, as plenty must have done throughout geological time. But prior to the Sun taking up a solitary lifestyle, the cluster in which it originated must have had billions of years in which to evolve. That was time aplenty for the formation of one generation of giant stars after another, all doomed to the same catastrophic demise.

Without such past goings on, our Solar System would contain no heavy elements at all. There would be no platinum, no uranium and no gold. The Sun, shining away as it fuses hydrogen in its main sequence, cannot make them. Without the helium fusion process that powered earlier generations of sun-sized stars through their red giant stages, there would be no significant amounts of carbon or oxygen. The Sun cannot, at least for the time being, make them in any great quantity.

No - it was the explosive death throes of earlier, now long-extinct star systems that made such elements available to the later generations. We are talking about elemental recycling on an unimaginably colossal scale. The downside, with respect to the bigger stars, is that their lifetimes were so short that the chances of the evolution of intelligent life, on any suitable planets that may have happened to orbit them, would have been pretty much eliminated.

Meanwhile, back on the beach

Talking of intelligent life, my rod-tip suddenly rattles in a determined manner. Fish on! I reel in, detecting a bit of weight at the other end of the line and in a few tens of seconds, a nice whiting lies pale on the sand. It, like me, is composed largely of light elements, organised into all sorts of complex molecular compounds. Carbon, hydrogen, oxygen, nitrogen (element 7), phosphorous (element 15) and sulphur (element 16) make up much of its flesh, while its bones are rich in calcium (element 20) and phosphorus. Iron, in the form of the compound haemoglobin, helps to transport oxygen to its tissues through its bloodstream. All of those elements, except hydrogen, owe their very existence to past nuclear fusion reactions, in the senior citizen years of stars that have long since departed the scene. From such stars were our atoms made and to interstellar space those same atoms will, billions of years hence, be returned. It's one way of looking at the long sought-after concept of immortality. The atoms making up that fish and its hunter alike will remain unchanged forever unless, perhaps, they eventually become part of another giant star that goes supernova at the end of its life.

Two whiting, filleted and fried in seasoned oatmeal, make for an excellent supper and I'm now halfway towards that goal. Having rebaited my hooks, I wade out into the shallows to cast. Putting the reel into open-spool mode, so that it can release that thin monofilament nylon line, I turn my body away to face back towards dry land. I then adjust the amount of line extending from my rod-tip down to the 140 gram streamlined fishing weight, made from lead, until "the drop", as we anglers call it, is just right. Holding the line tight against the body of the rod, I swing the lead weight away from me, landwards, then let it come back under the rod, at which point I twist, punch and pull forcefully in a rapid but smooth sequence of movements. The rod responds, compressing hard as I sweep its tip overhead, then straightening as I release the line. Immediately, the lead projectile hurtles up into the sky, towing the baited hooks and the line behind it, back towards deep space, the place of its origin. In just seconds, though, the momentum of the cast peters out, gravity takes over and the lead and baited hooks drop into the water, about a hundred metres out to sea.

None of that casting exercise, routine to any beach fisherman, would have been remotely possible were it not for those ancient supernovae or neutron star collisions, within which the heavy elements like lead were formed and distributed far and wide through the cosmos. Some five thousand million years ago, that lead now making up my fishing weight was scattered throughout our corner of the Milky Way galaxy. A few hundred million years later, its atoms were caught up, with countless others, in a vast, spinning nebular disc of matter, hundreds of millions of miles in diameter. At the heart of that spinning disc, there shone a new-born star. The star had no name back then. It would be another 4.56 billion years before we humans came into being and started giving things names: Helios, Sól, Surya, Ra or Sunnōn – take your pick.

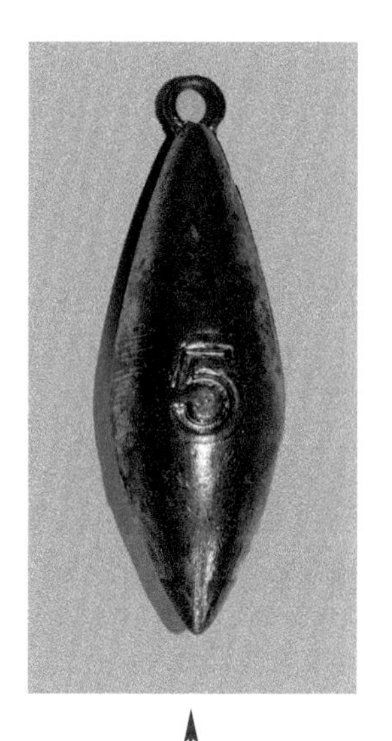

Element 82 - lead - making up a 140g fishing weight.

That spinning matter making up the nebular disc quickly began to clump together to form the orbiting bodies that were, in time, to become the planets, asteroids and comets of our Solar System, including our home, Earth. Initially, the lead was thinly distributed throughout our planet, but over the next few billion years, some of it was concentrated by geological processes, leading eventually to the formation of ore deposits. In turn, after another few hundred million years, humans discovered some of the ore deposits and mined them. They dug up the lead ore, smelted it into metal and made the lead into useful things, like fishing weights.

Contemplating this sequence of events, the gaze of the hunter on the beach returns to the Hunter in the sky. "None of this would have been possible without the likes of you and your ancestors", he muses to himself. "No lead weight, no rod nor reel, no fish, no beach, no Wales, no Earth." On that basis, existence is a precious thing indeed.

WALES THE MISSING YEARS

Part two: Origins
From insterstellar dust to Planet Earth

On a bright Saturday afternoon in the late 1960s, I was taken by my parents to a hill called Wren's Nest, looking out over the Black Country of the West Midlands. I recall the encounter vividly to this day. Here was a new and fascinating world of disused quarries, towering (to a four year old boy) crags, seemingly vast caverns, that I was told in no uncertain terms were, "very dangerous" - and fossils. Fossils were everywhere. Instinctively, I knew they were different. Strange, unfamiliar life-forms, set in stone: later I learned they were no longer to be encountered alive anywhere on Earth. They were remnants of the life that lived on Earth more than four hundred million years ago. A deep sense of awe and fascination was planted in my young mind, something that has never left me. I had entered the fourth dimension.

For all of us, there are only three dimensions within which controlled movement is possible: up or down, side to side and forwards or backwards. The fourth dimension, time, simply progresses along. But if you get into Earth Science, you can explore the fourth dimension, at least within your thoughts. It opens up a treasure house of discovery: the more you find out, the more you want to learn. Within the fourth dimension, there is a sense of liberation from the excesses of modern life with the ridiculous and unnecessary pressures we often inflict upon ourselves. For example: fashion, about which the fourth dimension does not care. That the fourth dimension also helps one better understand the planet, our only home, is a huge bonus, since if there's one thing we all need to do with the utmost urgency, it's that.

All sorts of people appreciate a trip into the fourth dimension. It's not a closed shop with a sign saying, "graduates only" stuck on the door: it's open to anyone who wants to take a look. For example, many people enjoy a good fossil hunt, hence the immense popularity of Dorset's Jurassic Coast.

Go there on a miserable winter's day and there will still be families and individuals scouring the foreshore to see what they can find – and some very important finds have been made by non-academics. Mary Anning (1799-1847) of Lyme Regis did just that, finding her first fossil icthyosaur skeleton well before she had turned twenty and following that discovery up with a string of others: plesiosaurs, a pterodactyl, prehistoric sharks and other fish, all at a time when women were excluded from professional academia. She ended up knowing more about the fauna preserved in the two hundred million year old rocks making up the local sea cliffs than did the gentlemen who purchased the fossils, wrote papers about them and often claimed the credit for their discovery.

What drives the instinctive urge to enter the fourth dimension? In Anning's case there was money involved, but I strongly doubt that was the sole reason. I suspect that in her case and in all others, the urge is also there and always has been because it satisfies our natural hunting instinct – and at that, Anning excelled. Hunting is something that evolution hard-wired into us, since at one time it was essential to our survival. Today, that same instinct is manifested in a far more diverse range of activities. Those, like me, who fish the coastal waters for their supper, are hunters, as are those who fall over one another in their frenzied contest for bargains in the January sales. Such apparently unrelated activities are just culturally different ways of responding to the same driving stimulus. The same instinct drives the collector of fossils or minerals just as much as it drives the gold prospector or mineral exploration-geologist. We are all hunters at heart.

Fossil Trilobites - no longer to be encountered alive anywhere on Earth.

Science in a nutshell

Scientific research again taps into the hunting instinct. There is the excitement of discovering something new; the satisfaction of following up a discovery and, eventually, explaining it by developing and presenting a hypothesis. That presentation is conventionally done in two ways: by giving illustrated talks to fellow researchers in the same discipline at seminars or conferences and by publishing a detailed account of your findings in the peer-reviewed scientific literature. Peer-review is a robust quality control filter. It does not catch every bad paper, but it gets most of them. If a bad one slips through the net and gets published, the scientific community will almost certainly notice and call it out, with a loss of reputation for its authors. When one considers the nonsensical pseudoscience circulating online, where there is often no such filter process, one can appreciate why peer-review is important.

A carefully considered and well presented hypothesis, supported by plenty of good old hard evidence, will gain increasing acceptance amongst other scientists working in the same specialised field. In such a way, a hypothesis graduates, in time, to become a theory. When a non-scientist says, "ah, but it's only a theory", they are missing the point. We do not talk of proof in the sciences: there are proofs in advanced mathematics and in legal jargon, but not in our case. Instead, we take available evidence and use it to explain a phenomenon. A hypothesis, then, is a relatively young but nevertheless credible explanation for something. It will be subjected to lots of testing so that it may either endure, or fall by the wayside. If it endures, it will eventually become a theory - a thoroughly investigated and independently replicated explanation for something, widely accepted by other scientists.

It's important to reiterate that by "other scientists", I do not mean a random mixture of them, but instead those specialists working in the relevant sub-discipline. One would not, for example, expect a particle-physicist to competently review a paper written by a geologist about a newly-discovered species of dinosaur. Neither would one ask a geologist to review an account of the latest findings from the Large Hadron Collider. It would be an unreasonable thing to request. To make objective comments on a piece of work, one needs specialised people who, through long experience in the same field of work, are capable of critically examining the presented evidence.

Hypotheses presented in the peer-reviewed literature share that information with an already engaged specialist scientific audience. The really difficult bit is getting the same information out to the wider public. That's not because the wider public are stupid – far from it – but because they may be inexperienced in the subject matter. I'm just the same when it comes to cars, to cite but one example out of many. I can see my car has a nasty case of corrosion when my mate gets it up on the ramps, but can I solve that problem? No, because I have no personal experience in welding. He can, however, expertly remove the corroded metal, weld a stout replacement into the hole thus created and the car thereby passes its MOT. Never underestimate experience. Without experienced people doing all the vital tasks for society to function, and without new people constantly entering occupations from welding to agriculture to the Earth sciences, all of equal importance, society would very quickly disintegrate. We all play our part.

Getting scientific findings out to a wider audience, as we are seeing with the current COVID-19 pandemic, can be of critical importance. As with climate change, we are up against people, some with a political axe to grind, some with an over-fertile imagination and others who are paid lobbyists, who deliberately create and disseminate misinformation about various science related topics. Imagine that latter one for a moment – being paid to mislead people. It doesn't float my boat. The fourth dimension is such an amazing place that there's absolutely no need to make up stuff about it.

Whilst geological academics work with the fourth dimension in the routine course of their research, there are very many people who seek out fossil or mineral specimens for aesthetic purposes, as objects to contemplate and appreciate, and why not? Well-crystallised minerals and well-preserved fossils are both beautiful and mysterious. Personally, in contemplative mood, I've always been especially attracted to rocks that have an unusual story to tell about themselves. This may be about how they were formed in the first place or how they came to be where they were found, or both. Once you get to know the fourth dimension, you start to develop the ability to read the evidence for such things, contained in the rocks, like the pages of a book.

A space rock from the deep past

In the context of rocks with tales to tell, one of my favourites is an expertly cut, polished and lightly acid-etched metallic meteorite slice, some fifteen centimetres in length and a few millimetres thick. It was prepared from one of several fragments found near a place called Seymchan in far eastern Russia. Just when the meteor blazed down through the atmosphere to impact Earth's surface is unknown, but in June 1967, geologist F.A. Mednikov discovered the first chunk of it, in the course of his routine surveying work. This initial find had a mass of 272 kilograms and was sat amongst the boulders in a dry riverbed. A second fragment of 51 kilograms was found nearby, four months later, showing that the meteor must have broken up before impact. In 2004, prospectors, armed with powerful metal detectors, located several more pieces. When found, these meteorites were a dull brown colour, just like any other lumps of iron would be if left in a stream bed for many years. It took painstaking laboratory work using specialist saws, polishing-laps and acid-baths to prepare the specimens in order to reveal their true magnificence.

Let's take a look at this space rock. Around its rim is a smooth, fused crust, where its outer surface melted as it sped down through our atmosphere, rapidly becoming an incandescent fireball, as meteors always do. But the inside is the really interesting bit. Much of the polished and etched surface of the meteorite slice consists of steely-looking metal. The etching has given the metal a distinctive criss-cross appearance. That's because there are in fact two different metals present, the one bright and shiny, the other rather dull, these being the alloys of iron and nickel known to mineralogists as kamacite and taenite. The two alloys dissolve at different rates when immersed in acid. Once the sawn slices have been polished, a quick etching in the acid-bath is all that is needed to bring out that criss-cross texture, known as the 'Widmanstätten pattern'. Widmanstätten was an Austrian count, one of the two people who discovered the texture, early in the 19th Century. In fact, English mineralogist William Thomson, living at the time in Naples, discovered the texture four years earlier in 1804, when acid-treating an iron meteorite in order to remove surface rust and he published his findings in a French journal. Disseminating scientific information in a Europe torn apart by conflict was not straightforward, though, so Widmanstätten, somewhat unfairly, got it named after him.

The Seymchan pallasite - a cut, polished and etched slice of this incredible space rock

Set within the crystalline metal groundmass of the Seymchan meteorite, there are scattered clusters of centimetre-sized rounded yellowish to orange-brown crystals of the mineral known as olivine. Olivine is a common silicate of iron and magnesium, found in abundance on Earth and, evidently, on other planets too: you may of heard of its green, gem-quality form known as peridot. Some samples of the Seymchan meteorite contain hardly any olivine; in others it's abundant. The distinctive and spectacular combination of metal and olivine places this meteorite within the class known as the 'pallasites', of which a number are known worldwide, each named after the place where it was found, as is the convention with all meteorites.

Critically, the formation of the Widmanstätten pattern, in which crystals of kamacite gradually separate out from taenite, can only take place during extremely slow cooling of the metal, after it solidifies from the molten state. In the case of the Seymchan pallasite, the cooling rate is estimated to have been just 7 °C per million years. Where could such a leisurely rate of cooling possibly take place? Surely few places could sustain such a process, but there is one that could - deep inside a planetary body, a location both hot and, just as importantly, well-insulated by its warm surroundings. Quite some journey, then, for that Seymchan space rock, from deep inside the heart of a planetary body, to its temporary place of rest in a Russian riverbed. The obvious question has to be, "how?"

For the answer to "how?", we must travel back in time to the very young Solar System, more than 4,500 million years ago. Back then, interplanetary violence was the natural order of things and it took place on a colossal scale. Planetary bodies, hundreds of metres to hundreds of kilometres in diameter, were chaotically orbiting the Sun in huge numbers and at great speeds. Some ended up on collision-courses with one another so that head on smashes were commonplace. Such high velocity impacts would often smash both bodies to smithereens. But go back just a few tens of millions of years and there were no planetary bodies at all, only the spinning, hundreds of millions of miles diameter, disc-shaped cloud of dust and gases that surrounded our newborn Sun. Let's start there, right at the beginning, and take a look at what we think happened.

Within the Solar disc, over a time thought to be no more than 100 million years, a remarkable transformation took place. The end product was a relatively ordered assortment of rocky planets like Earth or Mars, gas and ice giants like Jupiter or Neptune, various moons and lots of asteroids and comets. How this transformation came about is a complicated story, but one which science is, bit by bit, piecing together, so a summary may be given. There are plenty of potential sources of information. Meteorites, as we have already seen, contain useful evidence about the various physical environments in which they were formed. The Seymchan pallasite is just one of many kinds of meteorite: dozens of different types are recognised by the specialists who study and classify them.

Some meteorites are relics from that far off time when our Solar System was just beginning to develop - they represent the foundation stones of the place. In addition, technology now allows us to look so deeply into outer space that we are able to examine other young stars in unprecedented detail. We can see what's going on in their backyards. We have seen the great vulnerability of their discs to the intense radiation emitted by other nearby stars. Often, the discs are swept aside by solar winds and dispersed before any planets have had the time to form. In contrast, our young Solar System must have developed in a quieter neighbourhood. That kind of good fortune occurs, it is thought, with just ten percent of young stars. The Sun being one of that ten percent was the first of a series of critical events to pan out in our favour.

This is the sharpest image ever taken by ALMA. It shows the protoplanetary disc surrounding the young star HL Tauri. The substructures within the disc may show the possible positions of planets forming in the dark patches within the system.
Photo: ALMA (ESO/NAOJ/NRAO)

Planet birth

If you were able to travel back to the formation of the Solar System and scoop up a pint of the stuff making up that vast spinning disc-shaped cloud, what might you find? We can make an inspired guess, based on modern observations of other such clouds. Countless dust-grains, so fine that the naked eye would struggle to discern individual specks, in a gaseous, molecular soup of substances like hydrogen, helium, nitrogen, water, ammonia, hydrocarbons and carbon monoxide. The less common elements, including those heavier than iron, would be thinly dispersed throughout the mixture.

Our Solar disc was so hot at first that, within the zone closest to the Sun, many elements and compounds only existed as vapour. Further away from the Sun, where it became progressively cooler, increasing amounts of solids and liquids were present until a 'frost-line' was crossed beyond which water occurred in particles of its solid form - ice. As the whole system slowly cooled, chemical elements and compounds condensed from vapour into liquid then solid and at the same time the frost-line moved inwards towards its present position, inside the orbit of Jupiter. To this day, the frost-line marks an important divide, between the inner rocky planets and the outer gas and ice giants.

How the finely divided dusty solids of the Solar disc started to clump together into tangible objects is not so well understood, although that's clearly how the whole planet-forming process started off. Various hypotheses have been put forward and keenly debated, involving things like condensation of vapour around tiny dust particles and attraction between dust-grains due to static electricity. That uncertainty aside, it seems likely that clumps of dusty material quickly became super abundant. In addition, the contents of the inner Solar disc may have been modified by the Solar wind, always strongest nearer the Sun and causing light gases like hydrogen and helium to be swept out past the frost-line, leaving behind a predominance of rocky material. Once the 'planetary accretion' process, as it's known, was underway, growth was rapid. Collisions between the countless objects added to their size and in time compacted the dust into rocky bodies. Once above a certain diameter, the gravitational pulls of such bodies became appreciable, so that they began to attract all smaller objects in their orbital path, dragging material down onto their surfaces as they grew and grew. Planet formation was underway.

In order to better understand these early Solar System processes, it's good to have samples to examine - and we have, in the form of an especially interesting class of meteorites, known as the chondrites. Remnants of the most primitive planetary bodies of all, they have chemical compositions thought to reflect that of the original Solar disc. Chondrites are so-called because they are stuffed full of chondrules, these being distinctive, small rounded bodies, often less than a millimetre in diameter. Chondrules mostly consist of iron and magnesium-rich silicate minerals, such as our old friend olivine, accompanied by small quantities of oxides, sulphides and metals.

Slice of a chondrite meteorite, NWA 10256, 48x37mm
Photo: Christian Metz/Aerolite Meteorites

Radiometric dating of chondrules tells us that they formed over a short period of time, beginning some 4,565 million years ago. But chondrite meteorites also contain pale-coloured bodies known as Calcium-Aluminium Inclusions, or CAIs as they are often called since it's less of a mouthful. A few millimetres to over a centimetre in size, CAIs predominantly consist of calcium and aluminium-rich silicate and oxide minerals. Most CAIs that have been radiometrically dated are slightly older again, going back to 4,567 million years ago, making them the oldest known solid bodies to have formed in the young Solar System.

Chondrite meteorites also contain even older stuff that actually predates the formation of the Sun. 'Pre-solar grains', as they are called, are tiny particles of stardust a fraction of a micrometre in size, ejected from earlier stars as they died or exploded, billions of years before the birth of the Sun. Pre-solar grains include some pretty tough cookies like silicon carbide, which has a melting point of 2730°C, is almost as hard as diamond and is better known on Earth as the synthetic commercial abrasive, Carborundum.

Such pre-solar grains have been found in several meteorites, such as the Allende chondrite, a lump of space rock about the size of a large 4x4 that exploded over Chihuahua, Mexico, in February 1969. Several tonnes of bits fell to the ground near the village of Pueblito de Allende, over an area reckoned to measure some 50 x 8 kilometres. Pieces recovered from this 'strewnfield', as such areas are known, range in size from tiny chippings up to small boulders. The vast number of samples made available to science by this witnessed event has given the Allende chondrite the reputation of being the most studied meteorite in existence. Modern technology has made it possible to detect the ancient grains contained within such samples, thanks to their peculiar, pre-solar chemistry: they are quite clearly not of this Solar System.

How the chondrules were formed remains elusive. It has recently been proposed that they originated during frequent, ultra high speed impacts between the earliest formed rocky bodies, with the resulting heat generating and dispersing countless droplets of molten material. There are other hypotheses, too. Whatever the answer, the chondrules, CAIs and dust of the juvenile Solar System continued to accrete, eventually to form 'planetesimals' - or baby planets if you like. Many such objects were summarily destroyed in collisions while others spiralled down into the Sun, but plenty more survived. In their original form, the bodies from which the chondrite meteorites originated would have consisted of a mixture of CAIs and chondrules, gathered together in a rocky groundmass of compacted dust, like currants in a vastly over-sized bun. But things were not going to stay that way.

Meltdown!

As the planetesimals grew and grew, they became warmer and warmer, their temperatures partly raised by the transfer of heat energy from the continual impacts. However, there was another, far more important heat source. In the youthful and in any case very hot Solar disc, there also existed a great abundance of unstable, short-lived and fiercely radioactive isotopes, such as the positron-emitting aluminium-26. Such an abundance of radioisotopes can have few feasible explanations. One plausible source for so much unstable, radioactive material would have been a nearby giant star having gone to supernova – perhaps the same cataclysmic event that seeded the Solar disc with all those heavy elements.

Aluminium-26 is so radioactive that there's none of the primordial stock of the isotope left on Earth. It has all decayed into stable, non-radioactive magnesium-26. In the course of that very rapid decay process, emitting tremendous amounts of radiation, a vitally important thing happened. The heat literally cooked planetesimals from the inside. Not rare, nor medium, but well done - to such an extreme extent that larger bodies, containing more of the isotope, reached a completely molten state.

Why was that meltdown so important? The answer is that once planetesimals approached a molten state, the process of 'planetary differentiation' was able to commence. It's a vital part of our tale and it works as follows: in a part- to wholly-molten planetesimal, heavier elements such as iron tend to sink downwards towards the centre, through gravity. In doing so, they will displace lighter elements, such as silicon, which will tend to move upwards through the liquid, towards the surface. You can demonstrate the process in your kitchen, using household ingredients. Shake up a clear bottle containing olive oil and white wine vinegar and then let it stand. Over time, the less dense oil will work its way upwards to form an upper layer, with the denser vinegar sinking down to make up the layer below. The contents of the bottle have differentiated. Add a few of your favourite herbs and spices, shake again and hey presto! A tasty salad dressing.

Thanks to the planetary differentiation process, many planetesimals acquired an orderly, differentiated internal structure, consisting of a dense metallic core surrounded by a lighter, silicate-dominated mantle. We know this can happen to bodies down to quite a small size, too, because the asteroid Vesta, with a 500 kilometre average diameter, is also a differentiated body. In such planetesimals, if a total meltdown occurred it would have destroyed all the chondrules and CAIs: they would have become part of the melt. Paradoxically, though, we have numerous specimens of chondrite meteorites: they hit the Earth on a fairly regular basis.

Remember the large meteor that exploded over Chelyabinsk in Russia in February 2013? That was classified as a chondrite, once they'd picked the bits up and analysed them. It follows that some chondrule-rich planetesimals must have at least partly escaped the melting and planetary differentiation processes, either because they were too small in the first place, or because they broke up too soon due to impacts.

Conversely, the metal-rich meteorites, such as the Seymchan pallasite, clearly represent planetesimals that had already started to melt and differentiate, prior to their destruction in high speed collisions. That's why meteorites are so important: they literally track and trace events that befell the juvenile Solar System in those earliest years. They are the sole source of hard geological evidence from that time.

Formation of the highly distinctive metal and olivine mixture making up the pallasites is an interesting scientific problem for which several hypotheses have been advanced over the years. All involve destruction at a level difficult to imagine: compared to modern day disasters this stuff is off the scale.

One popular hypothesis proposes that the pallasites represent fragments from the core-mantle boundaries of differentiated planetesimals up to a few hundred kilometres in diameter, smashed to bits by the frequent high speed collisions that were a regular feature of the young Solar System. Such a boundary-region would presumably contain both silicates from the mantle and iron-nickel alloys from the core. Another line of thought suggests the pallasites came from planetesimals whose interiors were incompletely differentiated - a snapshot of the ongoing process, caught in the act as it were. Such hypotheses have an attractive simplicity to them since their requirements are at a glance straightforward. However, that doesn't necessarily mean they are correct.

Another pallasite-formation scenario, known as the 'hit and run' hypothesis, involves glancing impacts between differentiated planetesimals of certain specific sizes, one large and the other much smaller. Computer simulations of such collisions appear to show what happens. It's a bit shocking. Various forces are involved: not only shock but also shear and tidal forces due to the larger body's strong gravitational pull, all working on a mind-boggling scale. The silicate mantle and liquid metal outer core of the smaller body are literally ripped away, leaving behind the solid metal inner core. In turn,the ejected rocky and molten metallic debris coalesce together to form a string of new, smaller bodies, made up of mixed mantle and metal – pallasite through and through.

There is no strong scientific consensus yet as to which of these competing hypotheses of pallasite formation are correct. As meteorite specialist Peter Buseck wrote in a 1977 paper:

"If one were to write a geological Alice in Wonderland and try to think of the most handsome and yet improbable specimens, pallasite meteorites would be strong contenders."

One thing's for certain, though. It's impossible to explain the origin of space rocks like the Seymchan pallasite without invoking some kind of mechanism for forceful planetary disintegration. So, there you have it. One way or another, that meteorite specimen is a fragment of the deep inside of a smashed young planet. It's hard to think of a rock with a more violent history. Yet there it is, reclining peacefully upon my desk, next to a copy of the tide tables for Aberystwyth.

Following such planetesimal-disembowelling collisions all those billions of years ago, what happened next? This is an important question, sincc somcthing else clearly happened to create the metre-sized Seymchan projectile: it represents but a fragment of a once larger body of pallasite composition that itself had come from the aftermath of one of those old, giant impacts. There must once have been millions of such 'parent bodies', as they are called. Some of them must have struck other planetesimals in the crowded young Solar System, while others collided later on with the other rocky planets. Although we don't know the exact timing of the arrival on Earth of the Seymchan meteorite, the fact it was found on Earth's surface is a strong indicator that it was a relatively recent addition, for reasons we'll explore.

We'll come to what we know about the overall timing of events in a moment, but it seems clear that other bodies of pallasite composition took up residency in relatively stable orbits, such as within the inner part of the Asteroid Belt, that band of space rubble situated in between the orbits of Mars and Jupiter. Here there are hundreds of thousands of asteroids, the largest being Ceres, about 950 kilometres in diameter; the smallest being mere boulders, pebbles and sand grains. Although the Asteroid belt is not crowded as such - space probes can fly through it without too much difficulty - collisions must happen from time to time. In such instances, debris can get ejected in all directions, including towards the inner planets. It's thought that a large percentage of the various types of meteorites found on Earth had their origin in this region and manner.

This photo-mosaic synthesizes some of the best views the NASA spacecraft, Dawn, had of the giant asteroid Vesta, one of the largest in the Asteroid Belt with an average diameter of 525 kilometres. Dawn studied Vesta from July 2011 to September 2012. Intense impact-cratering reveals the nature of existence within this space rubble cluttered zone of the Solar System.

To get the Seymchan pallasite from the Asteroid Belt to Earth requires one such collision involving two objects, these being the parent body and another space rock big enough to do some serious damage. Some of the resulting fragments reach escape velocity, meaning their speed overcomes the parent body's gravitational pull, and off they go into space. At least one, a metre-sized chunk, is sent on a course towards the inner Solar System.

One of the marvels of modern chemistry is that we can figure out, through analyses, when this collision occurred, since any chunks ejected into space from a disembowelled asteroid are immediately exposed to galactic cosmic rays. That irradiation brings about subtle but detectable chemical changes. While deep within its parent body, such material is safely shielded from cosmic rays by its surroundings and once it lands on Earth, the planet's atmosphere and magnetic field likewise prevent that radiation from having any effect. The bit in between - the 'cosmic ray exposure age' as it's known - tells us approximately how long ago that final space journey began. In the case of the Seymchan pallasite, it was launched out from its parent body some 30 million years ago.

The metre-sized fragment of olivine, iron and nickel eventually took up an Earth-crossing orbit, meaning both bodies' paths crossed every so often, setting the scene for a potential impact. This is not uncommon - there are plenty such objects out there but the timing is critical in order for Earth and any such object to be in precisely the same bit of space at the same time. That's why larger impacts are relatively uncommon today. But this is exactly what happened with the Seymchan pallasite. Drifting lonely through the Solar System, one fateful day, 30 million years later, it found itself on a direct collision course with our planet. Encountering Earth's gravitational pull, down it came, blazing through our atmosphere, breaking explosively into several fragments, to land with a slap-bang-wallop on a mature planet teeming with advanced life. Its subsequent journey to the preparation laboratories - and in one instance onwards to my desk - was the easy bit.

Back in the young inner Solar System, as time went by, the gravitational pulls of the surviving and now respectably-sized planetesimals steadily cleared their orbital paths of much of the rocky debris. And so from chaos came order – with a system of four rocky planets: Mercury, Venus, Mars and – at almost the perfect distance from the Sun for its future biodiversity prospects – Earth.

WALES THE MISSING YEARS

Part three: Earth and Moon - The Hadean Eon
4,560 to 4,000 million years ago

The first full moon after the Autumnal Equinox is known as the Harvest Moon, or Lleuad Cynhaeaf in Welsh. Moonrise over the hill country of Mid-Wales can be a stunning sight. Dull yellow at first it creeps up slowly, its light bringing crests of distant ridges into sharp relief. Then, as it climbs higher and higher into the night sky it becomes a brilliant white disc, shining, magnificent, bathing the landscape in cold light. Even though it's more than 350,000 kilometres away, I can see many aspects of its alien geography with the naked eye. The whiteness is dappled with darker, slate-grey patches. With the assistance of a pair of binoculars, I can see mountainous tracts standing out proud around parts of its circumference. Craters are everywhere, some of them clearly enormous, with starburst-like rays of debris extending outwards for huge distances. Compared to Earth, the Moon seems to have had an extremely rough time of it, looking as if it has endured the cosmological equivalent to an artillery bombardment.

Both planetary bodies, in fact, have a lot more in common in terms of their early evolution than we once supposed, despite their strong visual differences. Only the surface of the Moon preserves so many physical features forged in the depths of ancient time: on Earth, such geological evidence has almost entirely been obscured. We'll come to the reason for that in a bit. But why do we have a satellite in the first place? Did it just happen along and glide nonchalantly into orbit around Earth, or did something more dramatic happen? The latter, it is thought. Of the various hypotheses, the most credible one, based on what we currently know, is that the Moon came into being within a few tens of millions of years of the Solar System's formation, when the young and still molten planet Earth was struck by a Mars-sized planetesimal, called Theia.

Let's try and put some context on an impact of that magnitude. Mars-sized means a planetesimal with a diameter of more than 6500 kilometres, considerably smaller than modern day Earth but still a substantial lump. Earth was also likely to have been somewhat smaller at the time of the impact, since collisions were still adding material to the inner Solar System planets. Purdue University, in collaboration with Imperial College London, has developed an online calculator to estimate the effects of impacts on Earth, so let's take a look at a few historic examples.

The Chelyabinsk impact of February 2013 is the best documented one of the lot. We don't need to run the impact calculator with this one, since there are plenty of videos to watch online. A twenty metre-wide asteroid streaks into our atmosphere, rapidly becomes incandescent with heat then blows itself apart, forming a blindingly bright fireball high over the Russian city. The estimated explosive yield is around 0.44 megatons of TNT. Thousands of windows are blown in along entire frontages, with colossal bangs as the shock waves arrive: dramatic stuff indeed.

On 30th June 1908, an even bigger asteroid (or possibly a comet) exploded high over Siberia. The Tunguska Event, as it's known, flattened millions of trees over more than 2,000 square kilometres. This airburst has been calculated to have involved a projectile some 60-100 metres in size, with an explosive yield in the multi-megaton range. In other words it was akin to the biggest historic nuclear detonations carried out by our military forces in their attempts to impress one other. Tunguska was the biggest documented event of its kind, with multiple eyewitness descriptions of the fireball and the shock wave blast. But let's ramp things up a bit more.

Carving out an 180 kilometre wide crater, the Chicxulub impact on Mexico's Yucatan Peninsula happened at the end of the Cretaceous Period, 66 million years ago. Widely blamed for bringing about the mass extinction that did away with the dinosaurs, the 17-kilometre diameter asteroid struck with an energy yield of around a hundred million megatons of TNT. This time, we'll need to use the impact calculator to have any idea what it was like. I entered the details based on an observer stood 1,000 kilometres away from ground zero. Here goes:

Within just over ten seconds after impact, thermal radiation from the fireball ignites our observer's clothing and nearby trees alike. After around three minutes, an earthquake occurs - one stronger than any in recorded history. Next, ejecta start falling from the sky and some tens of minutes after that, the air-blast arrives in the form of 900 mph winds, three times stronger than the strongest recorded tornadoes. A series of massive tsunami waves rake all coastal districts. It's pretty hard to imagine anything surviving that lot, even at 1,000 km distance.

When I entered a body with the approximate dimensions of Mars for the same location, I simply got this:

"Your position was inside the transient crater and ejected upon impact."

Ejected upon impact. Much of Earth had the same fate. The remains of the planet were engulfed in a vast, white hot cloud of rock vapour - not an elegant set of rings like those of Saturn, but a more dumpy affair shaped like a bagel or doughnut. As the cloud cooled, vapour condensed into droplets of magma. Depending where they condensed within the cloud, the droplets either rained back down to Earth or clumped together in orbit around the planet. The end product of this catastrophic event was a reformed Earth plus our familiar satellite, with a diameter of 3,475 kilometres, a mass of just 1.2% that of Earth and orbiting us at an average distance of some 385,000 kilometres. It's extraordinary that something so violent produced something so beautiful.

Details of the processes involved are still the subject of intense debate. The nature and dynamics of the debris cloud and the whole business of Moon-assembly are not yet precisely understood, especially since we've never had the chance to observe such goings on in the flesh. But one thing we do know. Moon rocks have strong chemical similarities, including matching isotopic ratios of certain elements, to the composition of Earth. We found that out by analysing samples collected on the geological field trips known as the Apollo Moon landings. The two bodies almost certainly came from the same source.

Such similarities throw up a problem, since physics predicts Theia to have been compositionally different to Earth. On a similar basis, we know Earth has differences in its rock chemistry to that of Mars, something we discovered by sending robotic landers bristling with scientific instruments down to the Red Planet's surface. The differences are not surprising because each planet likely formed in different regions of the young Solar System - with differing physical and chemical properties.

One possibility in addressing this problem is that the Moon-forming impact mixed Earth and Theia materials so thoroughly that they became compositionally homogenised. Objections to this idea do exist, but they may potentially be explained. For example, Earth's bulk chemical composition is thought to be enriched in iron, compared to the Moon. However, if Earth had already undergone planetary differentiation, prior to the Moon-forming impact, then a lot of its iron would already be gathered together in the planet's large, metallic core. If the impact left Earth's core undisturbed, but blasted away a large amount of the mantle, mixed with the vapourised remains of Theia, then anything formed from that resulting debris cloud would have the same, combined chemical fingerprint.

The iron discrepancy between Earth and the Moon in turn offers a clue to the size of Theia. Had Theia been a lot bigger, the impact would have resulted in core disruption, perhaps even leading to Earth's violent and complete disintegration into a new asteroid belt, occupying the region between Mars and Venus. On this occasion, we got lucky.

At first, the young Moon was a ball of magma. Planetary differentiation arranged its interior into a metallic core and a silicate mantle and once the lunar surface had cooled sufficiently, its magma oceans solidified to form a rocky crust, much of which has remained visible to this day. This original lunar crust, around 4.4 to 4.5 billion years old, makes up the Moon's brighter regions and it consists of a low density, whitish rock called anorthosite, containing an abundance of the pale-coloured calcium and aluminium silicate mineral, feldspar. Lunar anorthosites reflect a lot of the incoming sunlight from the Moon's surface straight back towards the Earth based observer. When the moon is full and the skies are clear, that light is bright enough to cast shadows.

Lunar sample 15415, collected by the Apollo 15 crew in 1971, is a chunk of anorthosite, some 10 cm across and composed mostly (98%) of the pale-coloured mineral, plagioclase feldspar. The crumbly look to some parts of the sample is testimony to the many impacts that have battered the Lunar surface in the time since its formation, although 15415 is a few hundred million years younger than some of the anorthosites making up the Lunar Highlands. However, at over 4 billion years old, it's still seriously ancient.
Photo: NASA

Although the anorthosites are bright to the eye, the brightest lunar features of all are certain impact craters, such as the prominent, 85 kilometre-wide, 4.8 kilometre-deep Tycho with its long, pale rays of ejected debris, down in the southern hemisphere. That extra brightness is because the debris was blasted out from several kilometres below the lunar surface, only an estimated 108 million years ago. It's relatively fresh. In contrast, the surface of the anorthosite crust was exposed to the ravages of the Solar wind, cosmic rays and a constant flux of micro-meteorites, a mixture of processes together known as space-weathering, for several billion years. Space-weathering leaves rocks looking a little duller, given sufficient time.

When Apollo astronauts first collected anorthosite samples, they found that much of the lunar surface consisted of a mixture of fine dust and broken rock fragments. Footprints left by the astronauts around their landing sites bear testimony to this. Such observations are unsurprising when one considers the great age of much of the lunar crust and the sheer amount of often overlapping impact craters with which it is pockmarked. Impacts shatter and pulverise rocks and the craters show, beyond any doubt, that the ancient anorthosites took one hell of a battering.

Craters within craters: a view looking down on India's Vikram Lunar landing site (image acquired before the landing attempt). Field of view: 87 kilometers across. →
Credits: NASA/Goddard/Arizona State University

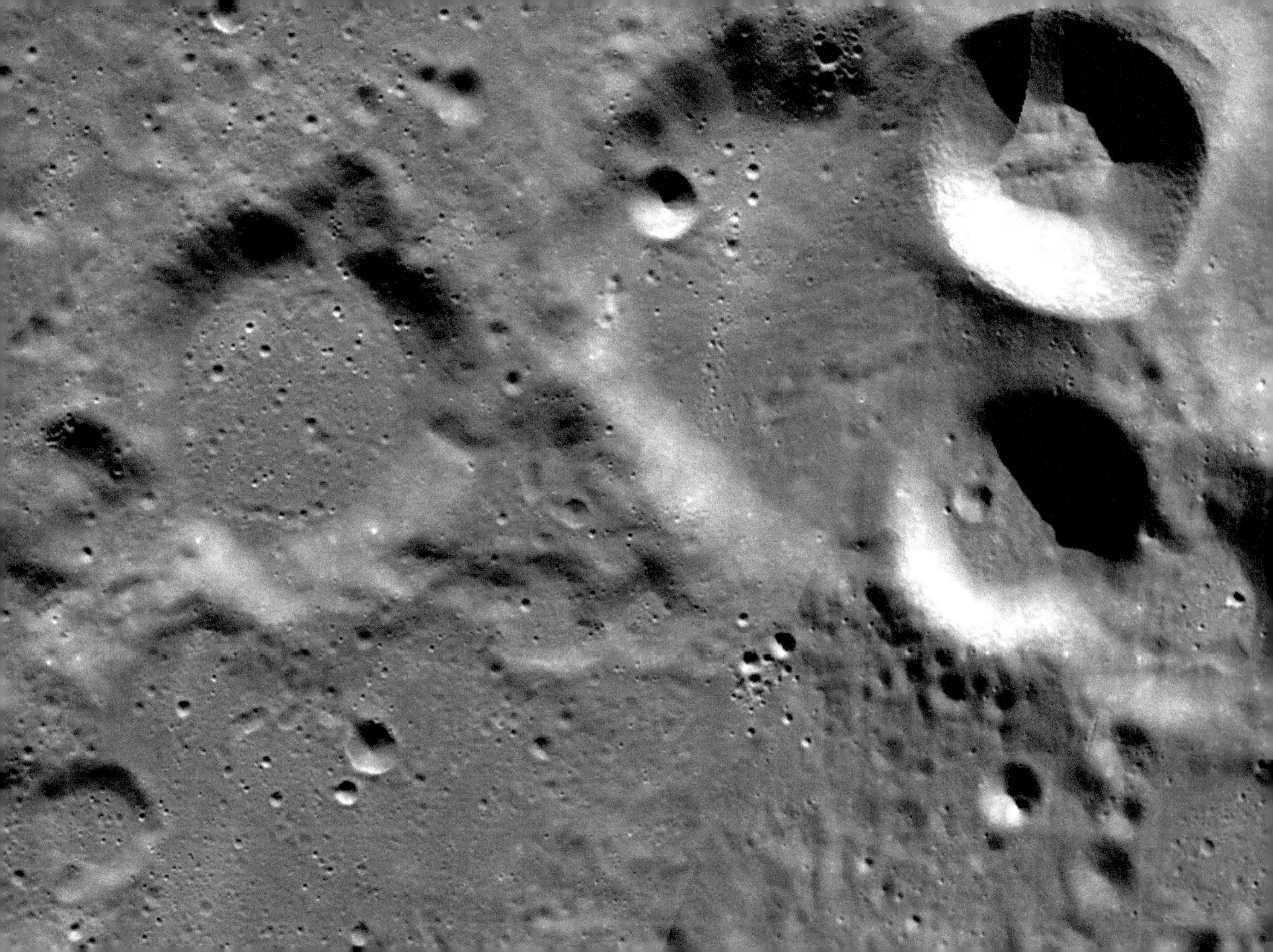

Seas of Magma

The darker parts of the Moon's surface have long been known as the Maria, from the Latin for seas, which is what early astronomers thought them to be. In fact, the Maria consist not of water but of vast sheets of basalt, a grey volcanic rock, rich in iron and magnesium silicates and erupted as great flows of lava. Unlike the classic volcanic cones familiar to all on Earth, lunar basalts typically erupted from fissures, often situated in the bottoms of huge impact craters, where the smashed up ancient anorthosite crust was weakest and most easily breached. Compared to Earth, the lavas were erupted in a low gravity environment, they were extremely runny and could flow over great distances. As a consequence they spread out over any low lying areas before cooling and solidifying.

Basalt is a dark grey to black, magnesium and iron rich igneous rock, meaning it has solidified from cooling magma. This is not a Lunar sample but is from Wales and is a great deal younger than anything on the Moon.

Radiometric dating of lunar basalt samples collected on the Apollo missions indicates that most of the volcanic activity took place between 4,000 and 3,000 million years ago. During this time, the Moon's interior was still sufficiently hot to produce copious amounts of magma. The volcanism subsequently waned and large scale eruptions had almost certainly ceased by around 1,000 million years ago. However, NASA's Lunar Reconnaissance Orbiter has provided something of a surprise in recent years, mapping localised features that are thought to represent much younger activity. Averaging just half a kilometre across, but scattered here and there across the lunar surface, they apparently mark the Moon's last gasp in volcanic terms.

The Maria are a lot less cratered, compared to the ancient bright anorthosites. That can only mean one thing: by the time of the most intense volcanic activity, the impacts that once so savagely battered the Moon had become less frequent. In turn, that means there must have been a span of time, prior to the Maria eruptions, when the inner Solar System was an especially dangerous place to be, a place where large, crater-forming impacts were commonplace – a different scenario to the present day.

Such a conclusion gave rise to the hypothesis of the Late Heavy Bombardment. This phenomenon, originally thought to have been at its worst from 4,100 to 3,900 million years ago, involved a sudden resurgence of asteroids and comets into the inner Solar System. However, recent work has challenged the hypothesis in terms of timing. Instead, the major period of impacts may have been much earlier in the history of the inner Solar System, so that by some 4,400 million years ago, it was all but done. But either way, the bombardment happened a long time ago.

Where did that influx of Earth-bound impactors originate? Many of the asteroids probably came from the Asteroid Belt, which was considerably wider and more cluttered with space rocks than it is today. As for the source of the comets, planetary scientists have looked out beyond the inner Solar System, to the distant, frozen regions. Here, the four gas and ice giants, Jupiter, Saturn, Uranus and Neptune, have their domains. Beyond the orbit of Neptune, vast numbers of planetesimals remain to this day, occupying a wide zone known as the Kuiper Belt. Pluto is one of many such objects.

Beyond the Kuiper Belt, there is the Oort Cloud, whose outer edge is about a hundred thousand Astronomical Units (AU) away. One AU is defined as the distance from Earth to the Sun, which is 149 million kilometres, so the outer Oort Cloud is a very long way indeed from Earth. These far distant, perpetually cold and dark outer zones of the Solar System are the source-regions for most comets - and there are an awful lot of them lurking out there.

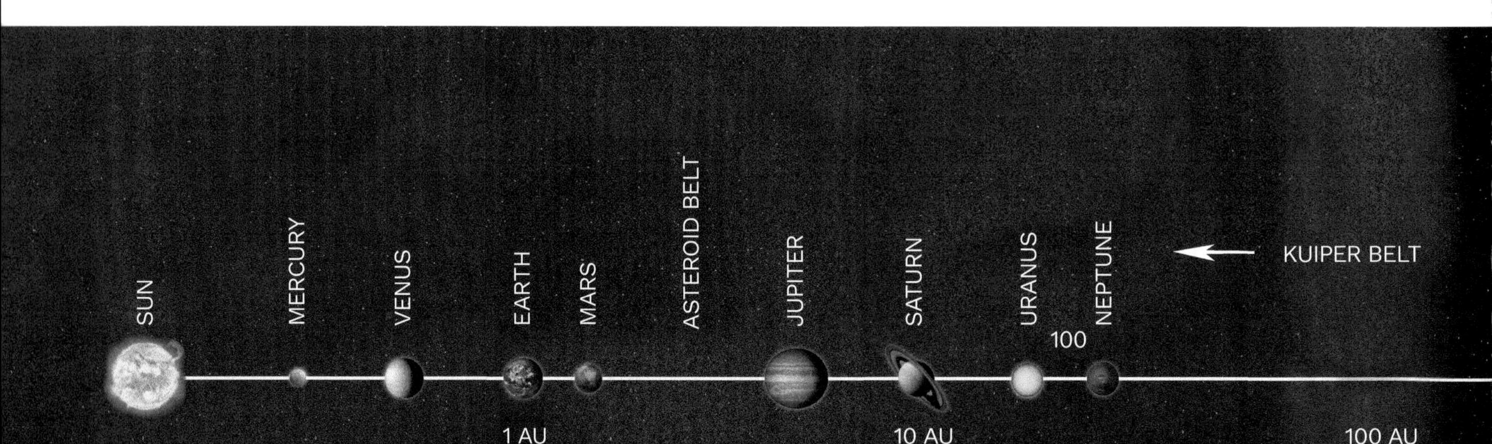

Computer simulations suggest that in the early Solar System, instabilities in the the orbits of Jupiter and Saturn were sufficient to disturb the orbital paths of Uranus and Neptune. The disturbances nudged the two outer planets still further outwards, away from the Sun, towards their current positions. In the process of such changes, countless other objects were encountered and scattered in all directions. Some were flung out towards the Oort Cloud, thereby disturbing the residents there. Others were deflected inwards, towards the Sun. The scattering seeded the inner Solar System with an abundance of potential impactors, many of which went on to bombard the rocky planets. That's why the lunar anorthosites are so heavily cratered, compared to the Maria basalts.

Although you can see the results of this space apocalypse by looking at the Moon through a pair of binoculars from your back garden, evidence for its effects is apparently lacking here on Earth. It's not that we missed out on the extravaganza: it's simply that on Earth, geological processes have obscured ancient impact structures, making them harder to find. It was only within the last few decades, after all, that the existence, in Mexico, of the relatively young end-Cretaceous Chicxulub impact crater was demonstrated. Compared to the Moon, Earth is a geologically hyperactive planet, meaning it is very good at obscuring evidence for events that took place in the deep past.

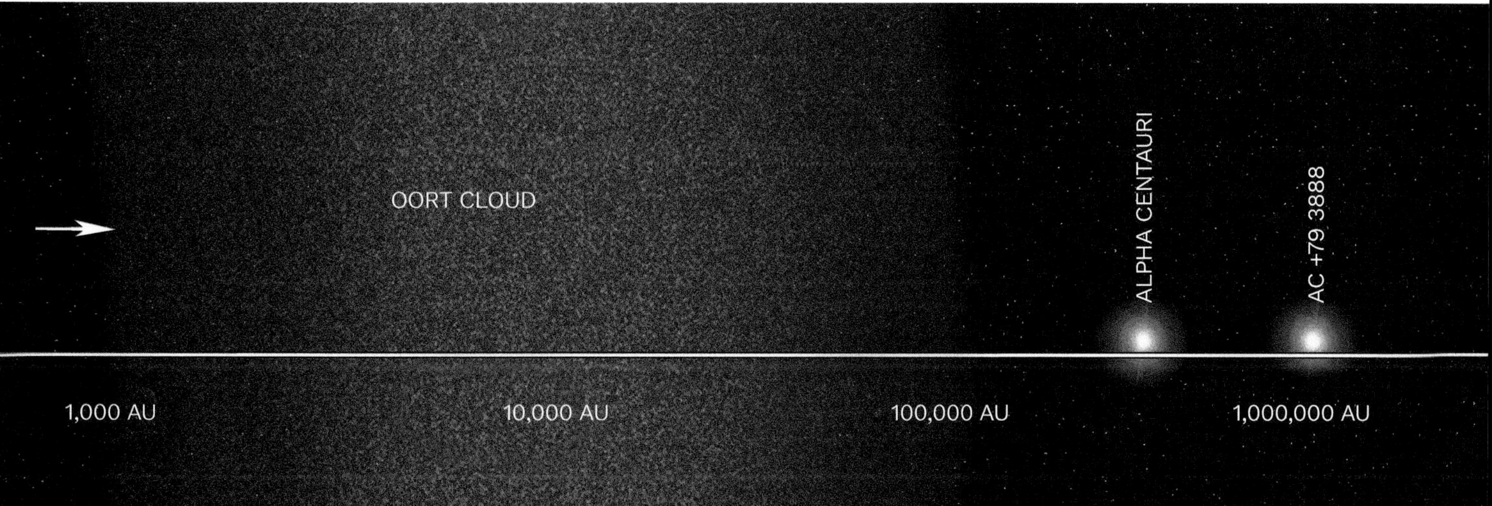

OORT CLOUD
ALPHA CENTAURI
AC +79 3888
1,000 AU
10,000 AU
100,000 AU
1,000,000 AU

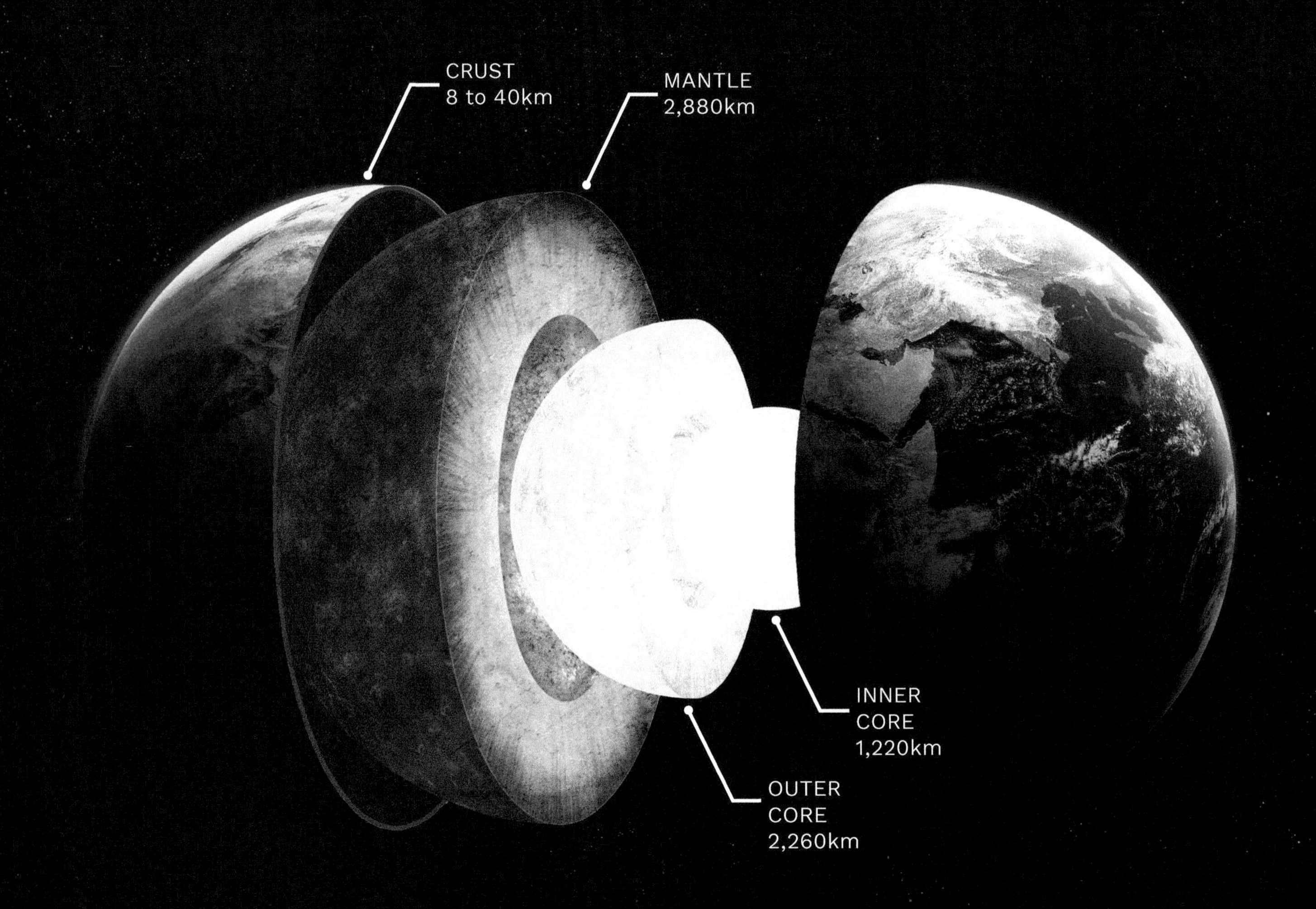

CRUST
8 to 40km
MANTLE
2,880km
INNER
CORE
1,220km
OUTER
CORE
2,260km

Architecture of a planet: Earth's vital magnetic field

Informally, the first 560 million years of Earth history are referred to as the Hadean Eon, named after Hades, the underworld of the ancient Greeks. During this time, the transition had taken place from a planetesimal through to a molten sphere undergoing differentiation. The Moon-forming impact disrupted things for a time but by the close of the Hadean, Earth was a well-organised planet, having an average radius of 6,371 kilometres, with a metallic core, a silicate mantle, a solid rocky surface, oceans and an atmosphere. Let's take a look at those key features.

The core of our planet has a radius of 3,480 kilometres and is dominated by the metals iron and to a lesser extent, nickel. It consists of two components, an inner core and an outer core. Their properties may be examined by the manner in which they interact with the seismic waves that are generated by earthquakes and recorded by geologists on their networks of seismographs. The seismic waves show that while the inner core is solid, the outer core, making up about two thirds of the total core radius, consists of liquid metal. Why the difference? It's mostly to do with pressure. Increased overhead pressure raises the melting point of most rocks and minerals and iron and nickel are no exception. Within Earth's inner core, the overhead pressure is a lot greater than in the outer core, because there's a lot more stuff sat on top of it. The outer core is only being sat on by the mantle, but the inner core has the combined weight of the mantle and the outer core to contend with.

Having a liquid outer core is a matter of the highest importance to the planet and its inhabitants. Electrical currents, generated by movements in the outer core's molten iron and nickel, maintain Earth's global magnetic field. Without a strong magnetic field, acting as a planetary shield extending well out into space, Earth's surface would have been constantly bombarded by the high energy particles originating from both the Solar wind and from cosmic rays. Such bombardment would have prevented the development of Earth's atmosphere and, in turn, the emergence of complex life. Recent work on ancient rocks in Western Australia has provided evidence that Earth had a strong magnetic field from early on, already acting to shield the planet and its atmosphere from those harmful particles, in what constituted yet another major slice of good fortune.

A different fate befell the atmosphere of our Solar System neighbour, Mars. Because Mars is a smaller planet, with a mass of around a tenth that of Earth, its gravity is weaker. Neither does it any more have a global magnetic field generated from its core: its smaller size meant that it cooled to the point that its spinning liquid outer core solidified. With that magnetic field generator stalled, the planet became fully exposed to the Solar wind. Although running water was once present in large amounts on the Martian surface, over billions of years the Solar wind progressively stripped the atmosphere away, leaving behind a cold and apparently lifeless world. As a consequence, Martian air is a hundred times thinner than that on Earth and consists almost entirely of carbon dioxide, with a bit of nitrogen, argon, oxygen, water vapour, methane and so on. Any water remaining on the planet's surface is present as ice.

Earth's outer core is surrounded by the mantle, some 2,880 kilometres in thickness and making up 84% of the planet's volume. It consists of dense silicate-dominated rocks rich in elements like iron and magnesium. Ambient pressures and temperatures are hard to imagine: in the lower, thicker part of the mantle which begins some 650 kilometres beneath Earth's surface, the overhead pressure is around 237,000 times the atmospheric pressure we experience here on Earth's surface. Temperature wise, the lowermost parts of the mantle near the mantle-core boundary are at around 4,000°C, whereas the uppermost parts are much cooler at just 500-900°C.

We know much more about the upper parts of the mantle because fragments of it get brought up to the surface in magma, thereby providing us with samples we can study. The geological term for such fragments is 'xenoliths' (pronounced zen-o-liths), a name derived from the ancient Greek and literally meaning 'strange rocks'. That's the only way to directly examine mantle material. We have yet to succeed in reaching it by drilling boreholes. The Russians did try to do just that, in the late 20th Century, with the Kola Superdeep Borehole, started in May 1970 and reaching an impressive 12,262 metres below surface. Unfortunately the project had to be abandoned in 1989 due to technical problems, but it's still the deepest man-made point on Earth.

Minerals present deep in Earth's mantle are often distinctly different from those that occur close to the surface, because of the higher temperatures and tremendous pressures that are ever-present down there.

One classic example is the element, carbon. When crystallised naturally in rocks close to Earth's surface, carbon typically occurs as the very soft, flaky dark grey mineral, graphite, from which pencil leads are made. Deep in the mantle, more than 150 kilometres down, carbon also occurs, but under such extreme conditions of pressure and temperature, it forms the transparent, crystalline substance that constitutes the hardest mineral of all, diamond.

The diamonds that are mined from various parts of the world are those that, just like the mantle xenoliths, have survived the rapid transit upwards in deep-formed magmas, in a situation akin to a geological high speed elevator. Upon cooling, such magmas solidify within their pipe-like conduits to form bodies of kimberlite and other uncommon, diamond-bearing igneous rocks that are mined in various places. We know that diamonds were incorporated into the kimberlite magma rather than forming from it. Chemical analyses and radiometric dating of tiny 'flaws' as gemmologists call them, these being inclusions of other mantle minerals trapped in some diamonds as they crystallised, have advanced our knowledge of the mantle by leaps and bounds. The dating shows that the diamonds are typically older than the kimberlite itself, sometimes by a factor of two billion years or more. That's something to ponder next time you're looking at a jeweller's display case.

An octahedral crystal of diamond, just 5mm across and complete with flaws, the little dark areas, which are the main focus of interest to geologists figuring out Earth's mantle.

Geologists have subdivided the uppermost few hundred kilometres of the mantle into two layers, based on important differences between their physical properties. The lower layer is known as the 'asthenosphere' and the upper one is the 'lithosphere'. While the lithosphere is the solid, brittle outer shell of the planet, the asthenosphere is ductile. A ductile substance is one that can be bent, stretched or twisted without breaking. Take a household candle as an example. If you try to bend one on a cold day, it will resist the pressure until it suddenly snaps in two. But leave an identical candle in a sunny place on a hot summer's afternoon and after an hour or two you can bend it without difficulty. The temperature increase in the candle wax has transformed it from a brittle substance into a ductile one.

A similar thing happens with rocks: heat them up enough and they will start to behave in a ductile manner. The rocks of the asthenosphere are able to flow very slowly, behaving over geological timescales as a sticky liquid, rather like the stiffer varieties of honey. As with the two components of Earth's core, the different physical properties of the lithosphere and asthenosphere can be spotted using seismology, because across the boundary there is an abrupt drop in the speed at which seismic waves travel.

The lithosphere is named after the ancient Greek word for rock, 'lithos'. Geologists have divided the lithosphere into two distinctive parts. Its uppermost part, referred to as Earth's crust, is the bit we live on, grow all our food on, drive on, clamber around on and extract minerals and fuels from. Earth's crust is relatively thin: compared to the mantle, it really is akin to the skin on a rice pudding with thicknesses varying, for reasons we will examine later, between a few kilometres and a few tens of kilometres.

As opposed to the distinction between the lithosphere and asthenosphere, with their differing physical properties, Earth's crust is distinguished from the rest of the lithosphere beneath it by having fundamentally important differences in its chemistry. The lower boundary of Earth's crust is thus a chemical, as opposed to a physical divide, with a switch beneath to lithosphere rocks with a mantle-like chemical composition.

On the modern Earth, we recognise two key types of crust, oceanic and continental. Oceanic crust, underlying all of our deep oceans as the name suggests, is typically just five to ten kilometres thick. Continental crust, making up the continents, with their plains, mountain ranges and the relatively shallow seas on their continental shelves, is more substantial with a thickness of thirty to seventy kilometres being characteristic. While continental crust tends to be dominated by pale-coloured, silica and aluminium-rich igneous rocks, such as granite, oceanic crust almost entirely consists of dark-coloured iron and magnesium-rich igneous rocks, such as basalt.

The physical processes that take place around the diverse topographic features of our continents have produced the bulk of Earth's sedimentary rocks. Sediments are generated by erosion and are transported by water, glacial ice, gravity and wind, to be deposited in lakes, rivers and their flood plains, estuaries, deltas and the sea. Chemical sediments, such as limestone, are deposited upon the continental shelves and sometimes in freshwater lakes. Evaporites, such as deposits of gypsum or rock-salt, can form when a body of salty water becomes isolated in a favourably arid climate and dries up.

In the Hadean, the composition of Earth's atmosphere, forming the planet's outermost layer, was very different to that of today - you wouldn't have been able to breathe the 'air' as such. However, by the end of the Hadean, a physical, chemical and geological framework had come into place that would allow Earth to go on to evolve into the planet we are now familiar with. Without that evolution, I could not have sat down and written this tale and none of you would now be reading it. But it was to be a long old haul. It would take up some 86% of the remainder of geological time between then and the present day, specifically from the end of the Hadean right through to almost the end of the Proterozoic Eon, and several more generous helpings of good fortune, in order to get there.

WALES THE MISSING YEARS

Part four: Introducing the Precambrian - Terraforming 4,000 to 541 million years ago

The Precambrian. Several decades ago, that's what they used to call what we now routinely refer to as the Hadean, Archean and Proterozoic. Back then, I was formally learning all about rocks, firstly at college in the Midlands doing A-levels under the well-remembered guidance of Derek ("Doc") Gobbett and then through a geology degree at Aberystwyth University, under Mike O'Hara and other leaders in their field. I will not forget any of them.

Once established in mid-Wales, I have to confess I spent an awful lot of time off-campus, for here was an expansive novel wilderness to explore, in the quiet hills of the Plynlimon massif and in the many old metal mines that riddle parts of the district. Underground or overground? It was often decided on the day. In the process, I met scores of interesting people, mostly of a non-conformist nature and with a common tendency to hang out in the front bar of the Angel – by far the best pub in Aber at the time.

Maybe such impulsivity is anathema to those who prefer pre-ordained structure in their lives, but to a rather awkward, shy 18 year old, the Welsh hills and their people were a wonderland. They were as much of a university to me as the concrete one atop Penglais Hill - and I am in no way decrying the latter. But there was something about rigidly conforming to the standards of the day that made me feel distinctly ill-at-ease. You went to college, graduated, maybe became a postgraduate, then either way off you went to the North Sea oil rigs, to earn lots of money and join the consumer society, like an item on a factory conveyor belt.

There was no mention of consequences. Today, with the knowledge instilled by reading reams of hard science in the research for this and other projects, I know that 'normal lifestyles' are in fact wrecking the only place we can call home. I was correct back then, but for all the wrong reasons.

Later, I went on to research the mineralogy of those same old metal mines, finding out new things about the ore-deposits and presenting the findings in a thesis, with the usual papers appearing in the peer-reviewed literature, conference presentations and so on. Being a postgraduate researcher showed me in no uncertain terms the importance of taking a strongly sceptical approach. By that I mean identifying any problems and looking at all available evidence to see if there were solutions. If that evidence was coming up short, I would go out again, to see if there was anything I'd overlooked on my previous field trips. I'm still at it today, since mineral deposits in Wales, with its highly diverse geology, are likewise highly varied and in some cases wonderfully problematic. They invite investigation.

The same point applies right across modern geology. If you cannot find obvious clues in the rocks, you start looking for less obvious ones. Advances in chemistry and physics - and the analytical techniques that have arisen from such things - have made it possible to examine samples in unprecedented detail, in ways that would have been considered the stuff of fantasy just fifty years ago.

I should at this point elaborate on scepticism. Scepticism is not just about identifying problems in your own work. It is also the vital process of questioning, testing and attempting to replicate the work of others - checking if their methods, observations, deductions and conclusions are correct. This is precisely how science progresses from one decade to the next, building upon what came before.

It is therefore unfortunate to have to record that in large parts of the media, the term, 'sceptic', is instead used to describe someone who is making a stand against an aspect of science that they - or their paymasters - have turned into a political football. Over a year into the SARS-COV2 pandemic, there are still people opining online that it is all a hoax. Such people, often making up or spreading conspiracy theories they find more appealing than the hard scientific facts, are not "sceptics". They are denialists - and regardless of whether their subject is rejection of the threat of SARS-COV2 or climate change, they should be completely ignored.

The Precambrian used to be regarded as a vast, impenetrably murky swathe of deep geological time. It was referred to by some in a derogatory manner at the time of my degree course – deformed, squished, cooked, even recrystallised rocks, regarded to be mostly devoid of interesting things one could write papers about, such as fossils. In that context, one of my former palaeontology lecturers, one of the best in the UK but not so keen on deformed stuff, used to refer to such rocks as 'twisted stones'.

Precambrian is a somewhat defunct term now as it probably deserves to be, given that it covers, like a dismissive wave of the arm, slightly more than 4,000 out of the 4,560 million years of Earth's story. Today, we conventionally divide geological time into four Eons: we have already looked at the informally named Hadean Eon, ending 4,000 million years ago. Next to come was the Archean Eon – the name meaning the 'earliest geological age', as that's what it was thought to be when it was so defined in 1872. The Archean lasted for 1,500 million years and was succeeded by the Proterozoic Eon, meaning the 'age of earlier life' and lasting for just over two billion years - a vast span of time. If, for the sake of argument, a person's typical life expectancy was 80 years, the Proterozoic would represent over 31 million consecutive human life spans.

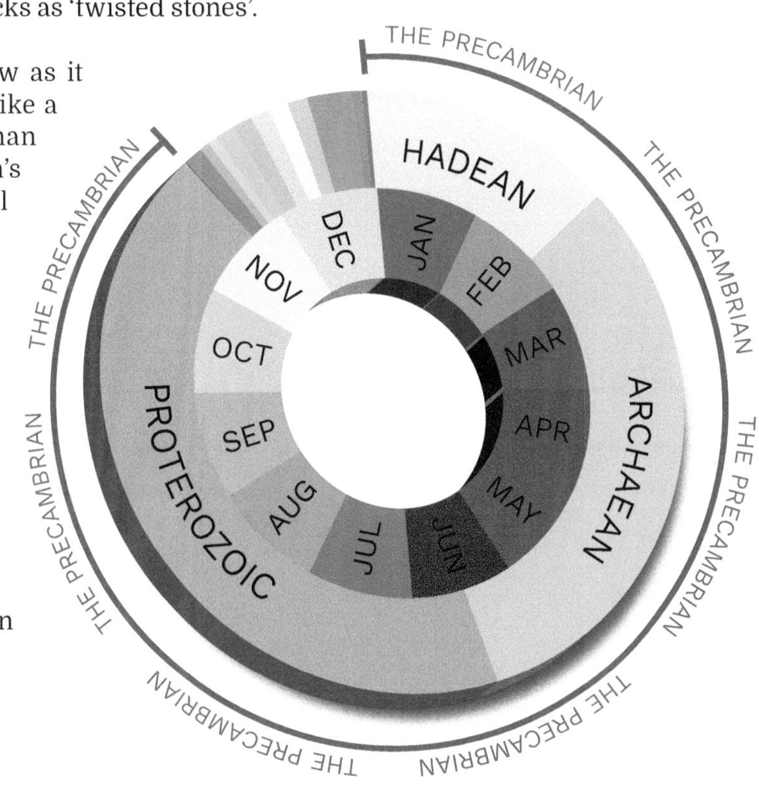

The Proterozoic – and indeed the Precambrian – came to an end 541 million years ago with the advent of the Cambrian Period. This was the first subdivision of the Palaeozoic Era of the Phanerozoic Eon - the latter meaning the 'age of visible life'. The Phanerozoic is ongoing today, so that its denizens include things like trilobites at one end and us lot at the other.

Hadean, Archean and Proterozoic: these three great divisions completely dominate Earth's history. Referring back to our plot of geological time set against the 12-month calendar, the three Eons take us all the way from New Year's day, through the rest of the winter, spring and summer and well into autumn, bringing us, at the start of the Phanerozoic, to approximately November 18th.

During the Archean and Proterozoic, the planet was, quite literally, terraformed. It was made a place fit to be inhabited by advanced life-forms. The conversion involved many natural processes, some operating over billions of years, others occurring much more quickly, but in combination, their consequences would be profound.

In the earliest Archean, Earth was a deeply hostile place; its environment was so unfriendly that had we been 'beamed down' to the planet's surface, we would not have lasted more than a few seconds before dropping dead. Yet by the early Phanerozoic, biodiversity was expanding into more and more habitats. What happened to make this possible?

The effects of the processes involved in this transformation are recorded in various ways in the planet's most ancient rocks. Understanding how the transformation happened, though, is an especially demanding area of research. The big difficulty in unravelling the deep past is due to the fact that, the older the rocks, the more likely it is that serious stuff has happened to them in the time since they were formed.

High-grade metamorphic rocks in a geologically famous road cutting in north-west Scotland.

Metamorphism

Serious geological disturbances, involving the effects imposed upon rocks by physical forces such as heat or pressure, are known collectively to geologists as 'metamorphism'. Metamorphic rocks are classified by the intensity of the metamorphism, from slight ('low-grade') to intense ('high-grade'). Low-grade metamorphic rocks may still retain original features that may be recognised and studied. In contrast, high-grade metamorphic rocks, heated to hundreds of degrees Celsius and subjected to immense pressures deep in Earth's crust, have in some cases literally been one step short of getting melted into magma. Like a wiped hard drive, such rocks have typically lost much of the geological information that was preserved within them, prior to the metamorphism, so their use in research has strict limitations. A lot of Earth's Archean rocks have suffered high-grade metamorphism - but not all of them.

Fortunately, there are places, scattered here and there around the planet, where we find outcrops of Archean rocks that have escaped the worst of those metamorphic changes. Original features can still be identified, examined and analysed, providing us with important information about the Archean environment. Such rocks are to be found in what geologists refer to as 'greenstone belts', named after their greenish colour that the relatively weak metamorphism has imposed upon them.

Greenstone belts typically form long, narrow zones, hence the 'belt' in the name. Such zones, consisting of volcanic and sedimentary rocks, are set within 'cratons', which are ancient and geologically stable blocks of continental crust. Representing the first small continents formed in the deep depths of time, Earth's cratons have in some cases remained relatively undisturbed for billions of years.

The volcanic component of greenstone belts often includes lavas that were erupted underwater: we can tell this because there is strong evidence for very rapid cooling, or quenching as it's termed. Studies have shown that the lavas solidified from magmas that, at the moment of eruption, were extremely runny and several hundred degrees hotter than any encountered on the modern day Earth. Such lavas are called Komatiites, named after the Komati River in South Africa, where they form part of the Barberton greenstone belt, one of several such belts set within the Kaapvaal Craton and some 120km long and 20-30km wide. In geology, it is standard practice to name rocks after where they were first discovered and described, this place henceforth being referred to as the 'type locality'.

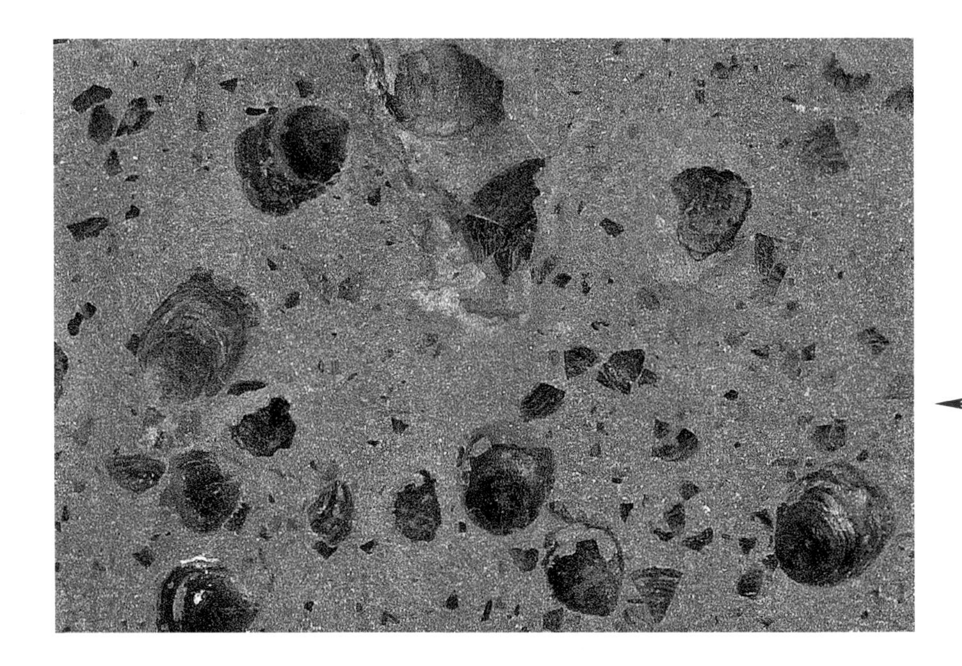

Slight metamorphism of rocks leaves most original features recognisable. This is Cambrian sandstone from North Wales and the fossilised shells of brachiopods are still visible, even though certain minerals within the rock have been altered or recrystallised.

At higher grades of metamorphism, rocks become completely recrystallised and new minerals are formed: original features become progressively obliterated. This rock was once a mudstone but has been converted into a mica-schist, with large embedded crystals of garnet. The sample is 18 centimetres across.

In the Kaapvaal Craton, the granitic rocks have a pedigree stretching all the way back to 3,700 million years ago. The Barberton greenstone belt is younger - but still well over three billion years old. Together, the craton and its greenstone belts are among the most studied Archean outcrops on Earth.

Komatiite lavas have a chemical composition close to that of Earth's mantle and they are dominated by iron- and magnesium-rich rock-forming minerals like olivine. Under a high-powered microscope, it is possible to examine the textures of and relationships between the component minerals making up a rock. This type of study, known as petrology, is enabled by carefully preparing sawn slices of rocks, grinding them evenly until the sample is just thirty microns (0.03 millimetres) thick. Such 'thin sections', as we call them, let the light through, so that different minerals can be identified by their physical, including optical properties, which differ from one mineral to another. The microscopic textures of komatiites are some of the most spectacular of any rocks on Earth.

The chemical composition of komatiites, the unusually high temperature of their parent magmas at eruption and their almost total restriction in occurrence to Archean greenstone belts, together serve to provide a geological hot topic. Just how were they formed in the first place? Many geologists have interpreted the physical and chemical properties of komatiites to indicate that Earth's mantle was itself hotter back then, compared to the present day. However, not everyone agrees with that, so there remains a lot of scientific debate to this day. Such observations, though, at least hint that the Archean Earth was a rather different planet compared to the one we are familiar with.

As science has progressed, we have come to understand many of the fundamental differences between the Archean Earth and that of today. By contrast, in the Nineteenth and early Twentieth Centuries, there were two broad camps of Earth scientists. On the one hand there were the Catastrophists, who advocated an Earth history involving long quiet periods punctuated by times in which disaster movie stuff happened. On the other there were the Uniformitarians, who thought that currently-observed, ongoing geological processes on today's Earth must have been responsible for all of the planet's geological features right from the start. It has now become increasingly evident that Earth's development involved a bit of both.

A thin section of komatiite lava under a petrological microscope, showing the bizarre skeletal crystal growths of olivine and pyroxene. Field of view: 5 mm

Photo: Alessandro Da Mommio of Milan

Water - an alien chemical?

Of the many changes that took place on Earth between the Hadean and Phanerozoic Eons, we'll start with one of the most basic of them all: the presence of water. Water is dihydrogen monoxide, as it can be called in technical terms or H_2O to give its chemical formula. Water. We relish a swig of the stuff on a hot summer's day or curse it if caught in a heavy rainstorm: it can create a colossal problem if it floods our homes and it can really mess with travel plans if it comes down in its solid form - snow. But in general, it's one of a number of everyday things all around us that we take for granted, mostly without a second thought.

Without water, life on Earth would not exist. Some critters were able to thrive in the oxygen-free atmosphere of the Archean, but without water, not a chance. That raises a rather interesting problem. Where did it all come from?

A rocky planet like Earth, formed at its distance from the Sun, in a hot nebular environment, ought to have been shortchanged with regard to volatiles, which are chemical elements and compounds with a relatively low boiling point. Water, which boils at 100°C, is just one example. Sulphur, carbon dioxide and halogens such as chlorine are others. Volatiles often cause explosive activity in modern volcanic eruptions, due to the sudden release of rapidly expanding gases. Deep in the Earth, such gases are contained within the magma by the great pressure down there. Upon the magma reaching the surface, a rapid depressurisation occurs and the volatiles immediately become very noticeable.

Prior to melting and differentiating into a core and mantle, Earth is thought to have had a bog standard chemical composition reflecting the nebular environment in which it had formed. In terms of meteorites, the closest compositional match would be that of the ordinary chondrite stony meteorites. However, the planet's content of volatile compounds is more akin to the fascinating carbonaceous chondrite bodies of the outer Asteroid Belt.

← Water: only when it means business do we even notice it.

Carbonaceous chondrites are a very special class of meteorite: their parent asteroid bodies were too small to have ever melted and differentiated. They formed in outer regions of the Asteroid Belt, in the near vicinity of the Solar System's frost-line. Samples of that part of the Solar System from primordial times, they are remarkably rich in water, reaching twenty percent by weight in some cases. The water is locked up in 'hydrated' (water-bearing) minerals. Carbonaceous chondrites also contain many other volatiles including organic chemical compounds.

A pristine, untouched meteorite, 30 mm across, sits on the surface of the Recovery ice sheet, near the Shackleton Mountains of Antarctica. This is one of dozens collected by the Lost Meteorites of Antarctica expeditions in the Antarctic summers of 2018-19 and 2019-20.

Image: Katherine Joy / The University of Manchester

In order to account for the amount of water and other volatiles on Earth, various hypotheses have been put forward and robustly debated over the years. The key question is this: was water always present on Earth from the time of its formation in the Solar Nebula, or was all or some of it introduced from an external source outside of Earth's orbit? The high volatile content of the planet suggests the latter. Potential external sources include comets and icy asteroids from beyond the Solar System's frost-line and those carbonaceous chondrites. How do we tell such potential sources apart from one another?

One method is to isolate ancient water from within a hydrated mineral in a meteorite sample, and measure its hydrogen isotope ratios. This is best done with witnessed meteorite falls, when samples can be recovered immediately afterwards, before natural processes on Earth's surface can cause any chemical changes such as corrosion. There's a better option, though. Go to Antarctica.

Antarctica is of course extremely cold but more importantly, parts of the continent are incredibly dry, technically classified as an icy desert. Corrosion and other forms of chemical alteration hardly happen, if at all, so the meteorites recovered from there tend to be absolutely pristine.

Over thousands of years, meteorites fall onto the ice and are carried along within it as the glaciers slowly flow down from the high ground of the interior towards the sea. But in places, such movement is impeded by coastal mountain ranges. Here, the ice flow stalls and the surface of the ice is blasted by the vicious winds that are such a big feature of the Antarctic climate. The wind is so strong it erodes away the ice from its surface downwards, leaving behind any rocks that were trapped within it. Teams of scientists scour such areas in the better weather conditions of the Antarctic summer and simply pick the meteorites up - there are several expeditions most years and a haul of a thousand meteorites per year is not uncommon. Antarctic meteorites are shipped back to research facilities, still frozen and in sealed clean containers to avoid any possible contamination. The information they yield when defrosted and analysed, in most cases, genuinely reflects that part of the Solar System environment from which they originated, including hydrogen isotope data with respect to any water they contain.

Broadly, hydrogen isotope studies work as follows: hydrogen has two key naturally-occurring stable isotopes. The common or garden version of hydrogen has one proton in its atomic nucleus, whereas the far scarcer and "heavy" version, deuterium, has one proton plus one neutron. The ratio of deuterium to hydrogen varies according to the water's original source in the Solar System, increasing in line with the outward distance from the Sun. Because the ratio can be calculated by carefully measuring the amounts of hydrogen and deuterium in a sample, we can therefore use it as a chemical fingerprint, pointing at whereabouts in the early Solar System that water originated.

Comets, from the frigid outer Solar System, have a range of hydrogen isotope ratios in their water ice. The ratios vary a bit from comet to comet, of those we've looked at in detail, in projects like the mission of the Rosetta spacecraft that made a rendezvous with Comet 67P/Churyumov–Gerasimenko in 2014. However, the ratios tend to be markedly high, compared to those present in the water of our oceans. Such results suggest comets were not the biggest contributor to Earth's store of water, although they must have played a part. Water from carbonaceous chondrites, on the other hand, has an isotopic chemistry that overlaps with oceanic water values. That suggests a potentially stronger link. I don't think we have solved this scientific problem beyond reasonable doubt, but we are certainly getting closer to doing so.

Carbonaceous chondrite asteroids were clearly caught up in the same events that we looked at earlier in terms of the massive cratering episode on the Moon in the Hadean. The disturbance, to put it mildly, crowded the inner Solar System with impactors. We're talking about billions of tonnes of asteroids a year, hitting Earth, its Moon and the other inner rocky planets - time and time again. That's potentially a lot of incoming water: even when vaporised during the impacts, it would end up in the atmosphere and in due course rain back down to the surface. It's also rather hard to ignore the fact that by the end of the most active phase of lunar cratering, Earth had liquid water oceans.

If the carbonaceous chondrite bombardment hypothesis is correct, and there is quite a lot of evidence in favour of that, then that event assumes critical importance. Had the barrage of asteroids not happened, Earth would have remained a dessicated, lifeless, inert world.

With far more certainty, geological evidence for liquid water on Earth is preserved in early Archean rocks whose features clearly show that they originated as sediments, transported and deposited in aquatic environments such as rivers, lakes or the sea. The oldest currently known water-deposited sedimentary rocks in the world are about 3,800 million years old and are to be found in the Isua greenstone belt in western Greenland. It was clearly raining by that time so streams, rivers, lakes and seas must have already been present. That said, depending on the composition of Earth's ancient atmosphere and therefore the rain, it might not have been the sort of stuff you'd want to water the plants with.

How do we know those oldest sedimentary rocks are a certain age? It's of paramount importance, in piecing together the way events played out in the deep past, to know what happened when. For that purpose, we use radiometric dating, a procedure developed in the earliest 1900s. Radiometric dating has come on in leaps and bounds in more recent decades as our analytical techniques have advanced.

One particular method of radiometric dating is especially reliable and will serve as an example. The technique involves an extremely useful mineral, zircon, which is zirconium silicate, $ZrSiO_4$. Zircon occurs widely as a minor 'accessory mineral' in many igneous rocks and is an extremely tough cookie by any standards, being very hard and with a seriously high melting point, way over that of any rock in the Earth's crust. So, once zircon crystals have formed, they are fairly difficult to get rid of.

Such incredible resilience is one part of the reason for the importance of zircon. The other part is down to the fact that zircon frequently contains traces of uranium, whose atoms fit neatly into its crystal structure. Uranium isotopes are radioactive: the relatively scarce U-235 has a half-life of about 700 million years and the far commoner U-238 has a half-life of about 4.5 billion years. Through radioactive decay, these isotopes slowly but steadily change, at constant fixed rates, into the stable isotopes of lead, Pb-207 and Pb-206 respectively.

The clever thing with zircons happens while they are crystallising, bobbing about in their parent magma. While a crystallising zircon happily accepts uranium as a trace element, it can't abide lead. For atoms of lead, there is a big No Entry sign. It follows that any lead present within a zircon was not originally there at the point of crystallisation, but has since formed, over geological time, as a product of the radioactive decay of those uranium atoms. Because we know the rate of decay and we know that doesn't change, we can figure out - from the amounts of those lead isotopes that we find - how long the process has been ongoing.

Radiometric dating of zircons therefore involves the painstaking analysis of individual, often tiny zircon crystals, carefully isolated in the laboratory from a powdered rock sample. The analyses determine how many uranium atoms are still there and how many daughter atoms of those lead isotopes are present. Those values in turn tell us how much uranium was there in the first place and the amount of time it has spent decaying, since it originally got trapped within the crystallising zircon. Since there are two different isotopes of uranium present, each with different but very specific decay rates, the analytical results can be cross-checked and each zircon crystal can be assigned an age.

Because of that very high melting point, older zircons may be naturally recycled. For example, if a rock gets melted, the resulting magma will contain that rock's zircon crystals. In many cases, due to such recycling, a rock will have populations of zircons with different ages. The youngest will be the ones that crystallised in the actual magma and the older ones will be recycling products from earlier geological events, representing rocks no longer with us, caught up long ago in destructive processes. Geologists have found examples of zircons, extracted from samples of somewhat younger Archean rocks, whose uranium-lead radiometric age is Hadean - older than any known rock on Earth.

Similar analyses can be done on zircons separated from sedimentary rocks such as sandstones. The zircons in this case will have been eroded from various older rocks and because they are so hard, they survive the erosion and transportation processes and get preserved in any sediment in which they accumulate. In a zircon population recovered from a sedimentary rock sample, the youngest ones will give us a maximum age of formation, since the rock cannot be older than the youngest zircons it contains. Few other minerals are so useful in this respect.

Regardless of the many scientific problems that concern this time, either being pursued hotly or remaining to get stuck into, we can now safely conclude that by the early Archean, Earth was in a reasonable shape to support some form of life, with liquid water - and therefore a tolerable temperature, neither too hot nor too cold. Just right. The planet had an atmosphere and a solid, rocky surface. One can see the potential for good things to come out of that combination. Now, we can explore the subsequent processes of terraforming in more and more detail.

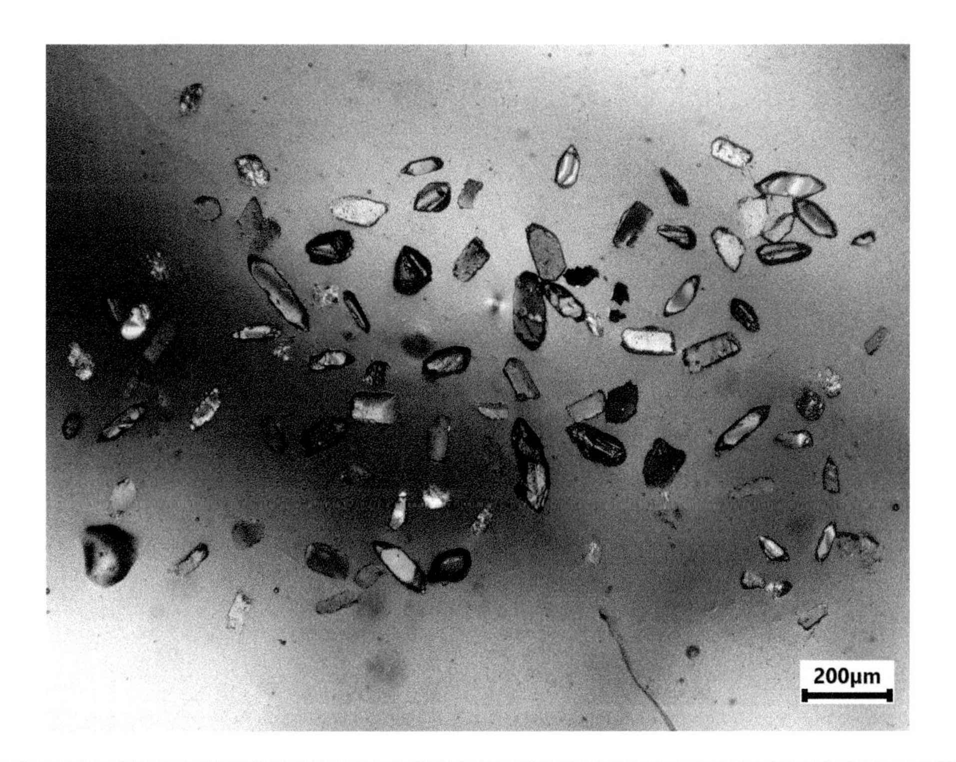

A population of tiny zircons isolated from a powdered rock sample, mounted in epoxy and polished and viewed under a polarising microscope. They are ready for specialist analyses.
Image: Dr Andreas Hahn.

WALES THE MISSING YEARS

Part five: Earth's greenhouse effect - a paradox solved

If we are starting to better understand what happened and when, what do we know about the Archean atmosphere? We know that free oxygen, forming almost twenty-one percent of the air we breathe today, was to all intents and purposes absent. We do have a good idea about what was present in the early Archean atmosphere, but we haven't yet nailed its composition in detail: we don't know so much about the proportions of the various gases that were present. Carbon dioxide, water vapour, nitrogen, hydrogen, ammonia, methane and other gases were all in the mix. Some of those gases have very useful properties indeed and there is plenty of evidence that they saved the planet from an icy, lifeless fate. How?

Back in the early Archean, the Sun's energy output was significantly less than that of the present day, since stable Sun-sized stars gradually brighten during their main sequence, as they work through their supply of hydrogen fuel. The brightness doesn't jump around everywhere. Instead, with stars of this type it's a slow but steady climb. Were it otherwise, Earth's climate state would have been perpetually chaotic, which would have completely messed with the progressive evolution of life.

The increase in solar luminosity, as it's known, has been about thirty percent, spread over the 4,567 million years since planets began to accrete together. Such a rate of change would be utterly imperceptible over human lifetimes, but over geological time it is obviously significant. In the early Archean, at the time that the oldest-known water-lain sedimentary rocks were being deposited, the Sun's strength would have been akin to an electric light with the dimmer-switch turned down more than a quarter - noticeably weaker, in other words. Under such circumstances, conditions should have been far too cold for liquid water to be present on the planet's surface, yet the place was clearly awash with the stuff. This, in a nutshell, is the scientific problem known as the 'faint young Sun paradox'.

Solving the faint young Sun paradox is an exercise that leads us straight to an essential property of Earth's atmosphere: the greenhouse effect. It's a term we tend to associate with carbon dioxide, but there are a number of greenhouse gases, some of which may have been highly important on the ancient Earth.

Firstly, let's chart the discovery of the phenomenon, a chronicle of scientific endeavour that takes us right back to the early Nineteenth Century. A few decades before that point, scientists had calculated the Earth-Sun distance, to within just four million kilometres of the modern, measured value of 149 million kilometres. That's important because it got people asking some major questions, beginning in the 1820s.

Jean Baptiste Joseph Fourier (1768–1830) was a French scientist who had previously undertaken big engineering and academic projects for the late Emperor Napoleon. With the Napoleonic Wars behind him, he turned his attention to investigating the physical world, with a special interest in the behaviour of heat. Fourier was the first investigator to work out that a planet the size of Earth, at its orbital distance from the Sun, should have been much colder than was the case. The average global temperature should have been significantly below freezing.

Fourier reasoned that there must be some other factor, apart from incoming solar radiation and geothermal heat, that kept the planet warmer. The suggestion he came up with was as follows: a lot of the incoming energy from the Sun, in the form of visible and ultraviolet light, that he termed 'luminous heat', was able to pass straight through Earth's atmosphere and heat up the planet's surface. In contrast, though, the 'non-luminous heat', now known as the infrared radiation that is given off by Earth's warmed surface, could not make the return journey unhindered. The air thus warmed by the 'non-luminous heat' must, he suggested, act as a kind of insulating blanket.

Unfortunately for Fourier, detailed physical measurements of the type required to further explore this hypothesis were not yet possible. Nevertheless, it later turned out that he had been extraordinarily close to the truth.

John Tyndall (1820-1893) was an Irish natural historian and pioneer in Alpine climbing. He was well aware that the scientific evidence available by the mid-19th Century indicated without doubt that, in geologically recent times, large parts of northern Europe had been covered by great ice sheets. However, what was much less clear was how the climate could have changed in such a drastic manner.

Picking up the baton from where Fourier left off, Tyndall considered whether such changes could have been caused by variations in the composition of Earth's atmosphere through time. Undertaking a series of experiments with various gases, he made the important discovery that water vapour was an effective heat-trapping agent. He also found that carbon dioxide was even better at trapping heat, despite being a trace gas occurring in the low hundreds of parts per million (ppm).

Hundreds of parts per million may not sound like a lot, but in fact many substances have important properties when present at such levels. For example, 500 ppm of hydrogen sulphide gas in air is a hazardous level, as any health and safety factsheet on the gas will tell you. It will make you unwell at that kind of concentration. Some other gases are extremely toxic at much lower levels, but let's not go there and instead consider the precious metals. To a prospector, thirty parts per million of gold in a rock sample would count as a very significant discovery. It's equal to thirty grams of gold to the ton of rock, or about an ounce per ton. That's easily ore-grade. Locate a few hundred thousand tonnes of ore like that and you've found yourself a profitable gold mine. These are all useful things to recall when someone dismissively tells you that carbon dioxide is "only a trace gas". It doesn't matter.

Tyndall's findings, with respect to water vapour and carbon dioxide, had in fact already been demonstrated independently, by American scientist Eunice Foote (1819-1888), in 1856, albeit using a different experimental method. In her account of the research, she stated:

"The highest effect of the sun's rays I have found to be in carbonic acid gas. An atmosphere of that gas would give to our earth a high temperature; and if as some suppose, at one period of its history, the air had mixed with it a larger proportion than at present, an increased temperature from its own action, as well as from increased weight, must have necessarily resulted."

Carbonic acid gas was the conventional name for carbon dioxide at the time. Once again, we run into the issue of who found what first, just as we did with the Widmanstätten pattern in iron meteorites, since the significance of Foote's work has only become fully appreciated in more recent years. She was the first scientist to point out the direct effect that carbon dioxide has on Earth's climate, a major scientific milestone.

It's also important that similar conclusions were reached, on different sides of the Atlantic, by two people working independently of one another, through different experiments. There is no evidence that Foote and Tyndall ever corresponded. Even looking at these results retrospectively, carbon dioxide and the other greenhouse gases were now firmly in the frame. Fourier, Foote and Tyndall really were the pioneers, but an awful lot more was to follow.

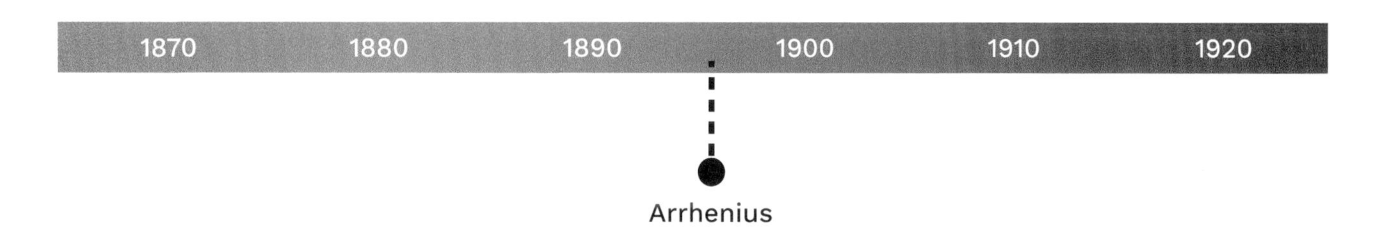

Even more weight was added to the importance of carbon dioxide at the end of the 19th Century, by Swedish scientists Svante Arrhenius (1859-1927) and Arvid Hogbom (1857-1940). Their research again involved ice ages, with Hogbom specialising in the behaviour of carbon dioxide in terms of the ways in which it is added to and removed from the atmosphere. A key paper by Arrhenius, "On the Influence of Carbonic Acid in the Air upon the Temperature of the Ground", published in 1896 and heavily quoting Hogbom's work, was the first attempt to quantify in detail the role of carbon dioxide as a greenhouse gas.

Arrhenius' paper also looked at longer-term changes to Earth's climate, going back tens of millions of years, and discussed the magnitude of the variation in CO_2 that might have brought about such changes. Although he did not specifically state in that paper that the burning of the fossil fuels would cause Earth's temperature to rise, Arrhenius would later go on to calculate the likely temperature effect that a doubling of its concentration might have. He came up with a value of 5-6°C of global warming. However, given the relatively low emissions of the time, he did not foresee an age when such a doubling might occur.

Further developments in this field of research took place throughout the first half of the 20th Century. In 1931, American physicist Edward Hulburt (1890-1982) ran calculations to determine the effect of doubling carbon dioxide once again, and, including the added burden of water vapour, he came up with a figure of around 4°C of warming. However, it was the post-Second World War period that really saw things take off. The Cold War was the key reason for the intensification of the effort. Atmospheric processes, such as the behaviour of infrared radiation, had major implications in military terms: for example, if missiles could be designed to home in on jet exhausts, these being infrared hotspots, they should be able to shoot down hostile aircraft.

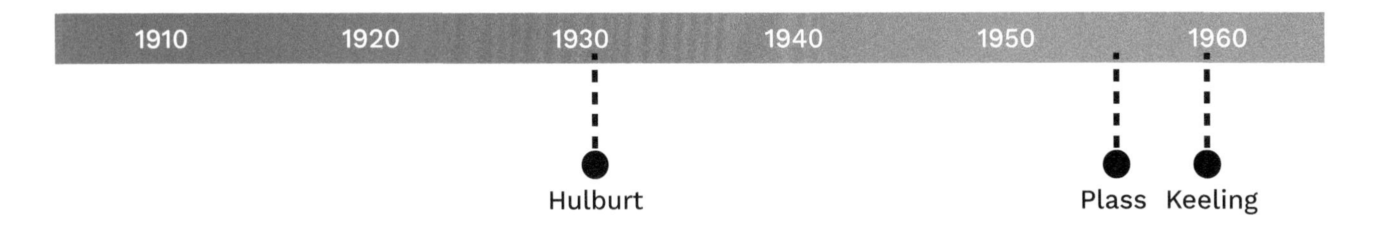

By the mid-1950s, scientists also had the tremendous advantage of the calculating power of computers. Such facilities made it possible to examine each layer of Earth's atmosphere and work out how it might absorb infrared radiation. Physicist Gilbert Plass headed up the task. His work, published in 1956 in the landmark paper entitled, "The Carbon Dioxide Theory of Climatic Change", confirmed that adding more carbon dioxide to Earth's atmosphere would without doubt have a warming effect. He calculated that a doubling of the atmospheric concentration of the gas would result in a global warming of 3-4°C. By the late 1950s, the work of Charles Keeling had improved the accuracy of monitoring atmospheric carbon dioxide levels and those measurements have continued to this day.

Although the terms global warming and climate change are often used interchangeably, the work of Plass demonstrated that increasing the carbon dioxide concentration of Earth's atmosphere causes global warming, which leads in turn to climate change. Likewise, reducing carbon dioxide levels leads to global cooling, leading again to climate change. It is the precise nature of the climate change that will vary in each of the two scenarios.

It is now more than sixty years since the Plass paper was published. Today, the consensus that increased greenhouse gas levels will lead to global warming and climate change is as solid as they come, among the thousands of specialists working in this area. It's become an extremely robust theory, supported by multiple lines of good old hard evidence. We know that greenhouse gases are those constituents of the atmosphere, both natural and man-made, that absorb and emit infrared radiation. When that infrared is re-emitted, it is radiated out in all directions, so that some escapes to space and some goes back down towards the surfaces of land and oceans, warming them and the lower atmosphere. This is Fourier's "insulating blanket".

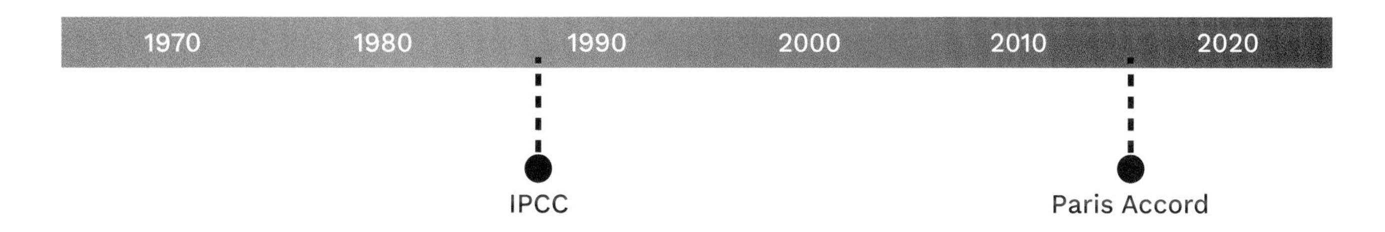

Earth's greenhouse effect thereby leads to a higher temperature at the planet's surface than would be the case if gases like carbon dioxide were absent. The greenhouse effect is only caused by certain gases: indeed, if the atmosphere simply consisted of nitrogen, oxygen and argon it would be essentially transparent to infrared radiation. In the absence of greenhouse gases, with modern solar output, the average global temperature would be a frigid -18°C. Instead, the figure is +15°C, and rising. Fourier was quite right to spot the discrepancy.

Like many good things, though, one can have too much of them. This is why climate change is currently becoming such a big problem. An analogy might involve malt whisky: a couple of measures may well be a pleasant way for you to round off a night at the pub and keep you hale and hearty for the walk home, but a whole bottle of the stuff consumed over a similar time span? Not so good.

Back in the early Archean, with solar luminosity at seventy percent of what we receive today, the planet should have cooled down from its hot, Hadean state, to become an iceball. Greenhouse gases stopped that from happening. It has been estimated that a thousand times more carbon dioxide than present day levels would have been needed in order to stop things from freezing up. This might seem an improbably high level, but perhaps the carbon dioxide was not acting on its own.

Some of the other gases present in the Archean atmosphere are known to have potent greenhouse properties, for example methane. As long ago as 1861, Tyndall had written,

"an almost inappreciable admixture of any of the hydrocarbon vapours would produce great effects on the terrestrial rays and produce corresponding changes of climate."

In the Archean atmosphere, containing negligible oxygen, methane levels may have been very high indeed and that gas is twenty times more potent as a warming agent, compared to carbon dioxide. Today's methane level of 1.875 ppm (and rising) is more characteristic of an atmosphere containing 21% oxygen, which is after all a highly reactive gas. In the unbreathable Archean air, hundreds - or even thousands - of ppm methane could easily have been present.

Whichever greenhouse gases were the most important will continue to attract research, but the bottom line is that the planet stayed mostly unfrozen. There is only one exception, this being a possible, apparently localised glaciation, whose distinctive sedimentary deposits have been described from mid-Archean rocks in South Africa. Otherwise, Earth was a comfortably warm place. Once again, a fortuitous combination of physics and chemistry had worked out in our favour.

What of our near-neighbours in the inner Solar System, Mars and Venus? We've explored both extensively with a variety of orbital and landing space probes, mapping surface features and looking at the chemistry of their atmospheres and surface rocks. It is abundantly clear that each planet experienced a very different evolution.

Mars has a very thin atmosphere, due as we have seen to the planet's global magnetic field having fizzled out at some point in the Archean when its liquid outer core solidified. With the Martian atmosphere largely stripped away by the Solar wind, its greenhouse effect was severely diminished, making it a predominantly cold planet.

Venus on the other hand, a planet only a little smaller than Earth, has experienced a runaway greenhouse effect. Its oceans have long since boiled away and with no means of sequestering carbon dioxide, the atmosphere is dominated (96%) by that gas. It spins very slowly - a day on Venus is 243 Earth days long and that slowness means there is no core-generated magnetic field although there is one in the upper atmosphere, due to the presence of lots of charged particles or ions. The Venusian magnetic field is weaker than that on Earth but it's still enough to protect the atmosphere from being stripped away as happened on Mars. However, that's not a lot of use when the surface temperature of the planet is over 460°C and the atmospheric pressure at surface is over ninety times that on Earth. Space probes don't tend to function for long even if they do make it down to a landing site on the Venusian surface.

If ever there was a Goldilocks planet, Earth had to be the one.

Part six: Life on Earth - when and how?

The possibility of a methane-rich Archean atmosphere is hinted at by the nature of some of the earliest known life-forms. It is thought that by the middle of the Archean, the class of microbes known as methanogens were thriving all around the planet. Methanogens absorb carbon dioxide and hydrogen as nutrients and then, within their cells, they produce water and methane. Importantly, methane is expelled into the atmosphere. Still occurring on today's Earth, their varied range of habitats includes the intestines of people, where they are responsible for the methane expelled from the bowel during flatulence.

Consideration of methanogens brings us onto the fascinating topic of the origin of life on Earth. By the late Archean, anaerobic microbes, plus blue-green algae known as cyanobacteria, were widespread in the planet's seas. But tracking down the first appearance of living things, much further back along the timeline, has proved elusive - and not for want of trying. Such elusiveness should hardly come as a surprise, since determining whether a microscopic blob, contained in an ancient rock sample, represents something that was once alive is a fiendishly difficult task. Indeed, the prevailing view for decades, ever since Charles Darwin suggested it in 1859, was that, "no fossil wholly soft may be preserved".

If Darwin was correct, given that early life was entirely soft-bodied, there should be no trace of it in the rocks. However, we now know that under certain circumstances, organic matter can be preserved, for example by becoming mineralised by durable substances such as silica and thereby sealed off from the outside world. Not only that, but life can also leave behind distinctive chemical 'biomarkers' in the rocks, calling cards from deep time, that can be detected by modern analytical techniques.

One such biomarker is the stable isotope, carbon-12, which all living things take in preferentially to carbon-13. By analysing how much of both isotopes are present in a carbon-bearing sample and calculating the ratio of carbon-12 to carbon-13, information can be obtained as to whether the sample had a life-related (or 'biogenic') origin: high ratios indicate the latter whereas low ratios suggest an inorganic source.

In the ongoing search for ancient life, much attention has been focussed on the Pilbara Craton in Western Australia. The craton contains several greenstone belts, the sedimentary rocks of which include a particularly important unit, the Strelley Pool Formation, 3,426 to 3,350 million years old and featuring sandstones, limestones and silica-rich rocks known as cherts, all originally deposited in shallow water. Within such rocks, there are abundant stromatolites, these being distinctive, multi-layered mound-like features that were once living microbial colonies. Such colonies have persisted right through the fossil record and can be found in warm, shallow water habitats today, so we can see what makes them tick.

Stromatolite from the Strelley Pool Formation of Archean age, Pilbara Cration, Western Australia.
Image: James St. John.
Specimen: Cranbrook Institute of Science collection, Bloomfield Hills, Michigan, USA.

Modern stromatolites are made up from extensive, thin slimy films of micro-organisms that are sticky, so they trap small grains of sediment. As the films silt over, those microbes repeatedly work their way back up to the surface, where they multiply and spread out to form another film or mat. Over time, with multiple silting and mat-forming episodes, a series of layers slowly builds up into a mound. The resemblance of modern stromatolites to layered features preserved in ancient rocks such as the Strelley Pool Formation is so remarkable that it is attractive to conclude that they must record ancient life. But it's not quite that simple.

Unfortunately, very similar, stromatolite-like structures can also form through sedimentary and chemical processes within evaporites – deposits of minerals like rock salt or gypsum left behind by evaporating bodies of highly saline water. Hot springs, gurgling out hydrothermal fluids, these being warm to boiling hot groundwaters laden with dissolved silica and other substances, can likewise produce similar layered features. The Strelley Pool Formation also includes evaporites and hot spring deposits. It would, wouldn't it?

For the geologists who work on such rocks, the hunt for ancient life has been greatly facilitated by decades of innovation in microscopy. Modern electron microscopy techniques have permitted the detailed, three-dimensional examination and chemical analysis of tiny cell-like structures in ancient rock samples. Sometimes, such structures turn out to be not what they seem. They may instead be deposits of an inorganic mineral whose crystals just happen to resemble something that might have once been alive. However, some of the features from the Strelley Pool Formation are more convincing. There is carbon-rich material present with a high carbon isotope ratio, plus other geochemical biomarkers and structures that resemble cells, some in apparent colonies.

Some thirty kilometres away from Strelley Pool is the outcrop of another ancient sedimentary rock, the Apex Chert, a bit older again at 3.46 billion years. Recent investigations, carried out on samples of Apex Chert collected back in the 1980s, have yielded interesting results. This analytical work has been done using spectroscopic technology that was only developed over the past ten years. Results of research by one team, published in late 2017, describe a surprisingly complex community of ancient microbes, each with its own apparently distinctive chemical fingerprint. There are said to be two different kinds of phototrophs - critters that made energy from sunlight, a methanogen and two types of methane-consuming bugs.

The discoveries in rocks like the Strelley Pool Formation and the Apex Chert are not free from controversy, with other researchers contending that the structures may be of inorganic origin. The problem is simple: the forms exhibited by potentially ancient microbes in these very old rocks – spheroids and ovoids, filaments and films – are difficult to distinguish from similar shapes produced by non-biological processes. Yet the controversy itself has served to further refine the criteria used to distinguish the remnants of early life from features of inorganic origin. Those developments are important too in the search for traces of life on other worlds: we have to be certain of what we're looking at, whether in the Archean rocks on Earth or, for example, on Mars.

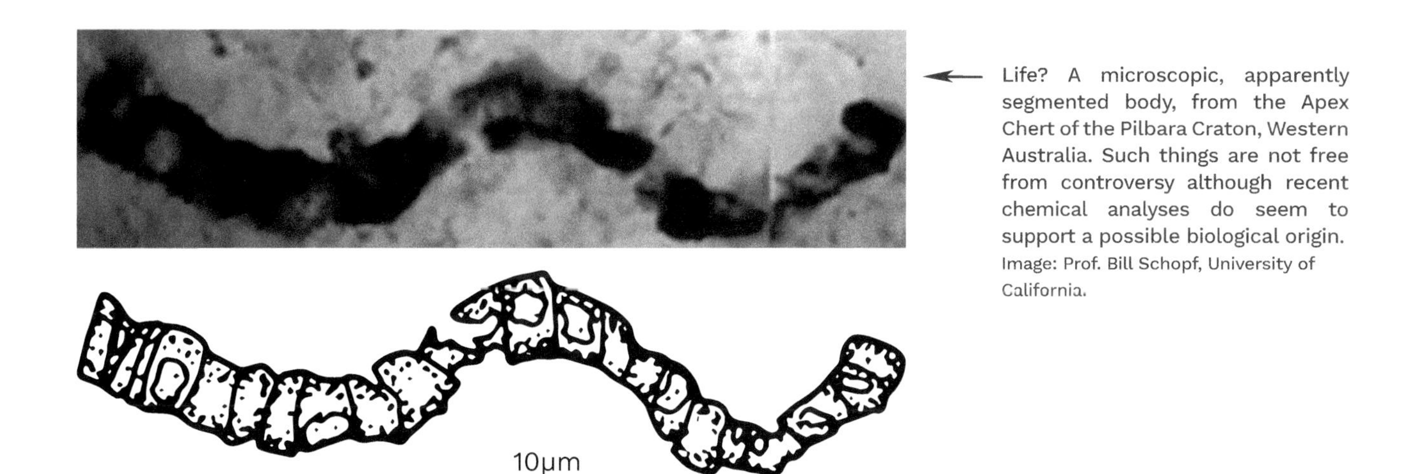

Life? A microscopic, apparently segmented body, from the Apex Chert of the Pilbara Craton, Western Australia. Such things are not free from controversy although recent chemical analyses do seem to support a possible biological origin. Image: Prof. Bill Schopf, University of California.

Once the various controversies are resolved, there is still the question: if life was well-established by some point in the Archean, it must have originally appeared still further back in geological time, so when? Only recently, sedimentary rocks have been described from the Isua greenstone belt in Greenland, dating back to 3,700 million years ago and containing structures resembling stromatolites. Again, these are of controversial origin.

In Quebec, Canada, strange filament-like structures, occurring in iron-rich precipitates from ancient, underwater hydrothermal vents, have recently been described in the scientific journal Nature. The iron-rich rocks date from 3.77-4.28 billion years ago, the implication being that life may have appeared as far back as the late Hadean. Indeed, in a recent paper on the early Earth, the authors included the following sentence, written in a remarkably matter-of-fact style, when one considers its implications:

"The estimated median age of the last impact big enough to vaporise the entire ocean is ~4.3Ga ago (4,300 million years ago), which provides a crude upper age limit on the origin of life."

How life actually formed in the first place is one of the most fundamental questions of all. But we understand parts of this puzzle reasonably well. The chemical building blocks of life known as amino acids were probably widespread on Earth from early on. By the start of the Archean, as we have seen, Earth was certainly habitable by primitive life, with liquid water oceans, an atmosphere, tolerable temperatures thanks to the greenhouse effect and a strong enough magnetic field to shield the surface from harmful cosmic rays and the solar wind.

The Hadean bombardment of Earth by carbonaceous chondrite asteroids and meteorites brought more than just water to the planet. Analyses of such meteorites have shown them to contain a wide range of organic compounds, including amino acids. We also know that comets, as recently demonstrated by the audacious landing of the Philae probe on Comet 67P/Churyumov-Gerasimenko, can contain a remarkable hotch-potch of organic chemicals. Instruments aboard Philae detected sixteen different organic compounds, including familiar ones like acetone and others with more non-household names, like propionaldehyde.

Recipe for amino acid soup

In addition to the discoveries involving meteorites and comets, we have known since the 1950s that amino acids are not difficult to make, thanks to the Miller-Urey experiment. This experiment, named after the two American scientists who carried it out, involved sealing a mixture of methane, ammonia and hydrogen in a five-litre sterilised glass flask, connected to a second half-litre flask of water. Heating the water in the smaller flask produced steam, which was then able to mix with the other gases. Continuous high-voltage electrical discharges were then sparked back-and-forth through the mixture, simulating lightning strikes.

After a week of such zapping, the whole system was cooled down again, whereupon the water vapour condensed back into liquid and was collected. It had turned a pinkish colour. Great care was taken to isolate the water from outside contamination such as airborne microbes, prior to examining its chemical properties. Several amino acids were found to be present. Many variants of this experiment were subsequently performed using additional gases, such as carbon dioxide and hydrogen sulphide, in many cases producing various complex organic compounds.

Just how such molecules got to the point where they were able to multiply and spread around is a question that has yet to be answered. But by the end of the Archean, simple, single-celled water-dwelling life was widespread. One of the most important groups were the cyanobacteria, simple blue-green single-celled or filament-forming microbes that are still abundant on Earth today, in pretty much every imaginable ecological niche. That's some success story. Cyanobacteria were to bring about an especially fundamental change to the planet through the process known as photosynthesis, by which plants absorb carbon dioxide from the atmosphere and release oxygen in its place.

Photosynthesising microbial life was therefore an oxygen source: atmospheric gases have both sources, from which they become freed, and sinks, into which they become locked up. Both the freeing and locking up result from chemical reactions: for example, metallic iron readily reacts with atmospheric oxygen in the presence of moisture to form rust, which is iron oxide, a relatively stable substance and therefore an example of a long-term oxygen sink.

In the Archean, the negligible free oxygen in Earth's atmosphere would have prevented that rust-forming reaction from taking place. So what happened to the iron instead? Consider: it's a fairly abundant element and common igneous rocks like basalts often contain several percent of iron, bound up in various minerals. Basalts have always been prone to chemical weathering – involving the reactions between corrosive substances dissolved in rainwater and those iron-bearing minerals.

In the absence of free oxygen, all of that iron was simply dissolved and carried off in solution, down into the rivers and on into the seas. Likewise, iron-rich waters were being discharged from seafloor hot springs in areas of volcanic activity, the iron remaining in solution. The overall result was that the world's oceans were, at the time, anoxic and iron-rich, compared with the modern day iron-poor, oxygenated waters.

Then along came the cyanobacteria. Once these photosynthesising critters became abundant and widespread, copious quantities of free oxygen were being discharged by them, as tiny bubbles, into the seawater. The oxygen reacted with the dissolved iron and dropped it out of solution, precipitating it onto the seafloor as red and black iron oxides in layers alternating with silica. Such 'banded iron formations', as they are known, accumulated in abundance on sea floors around the world: some are hundreds of metres in thickness. They are of widespread occurrence within Archean and, to a lesser extent, Proterozoic sedimentary rock sequences, constituting a major modern source of iron-ore.

The distinctive banding that gives banded iron formation its name seems likely to represent periodic changes in ocean chemistry involving the availability of both iron and oxygen. More importantly, these extensive Archean iron deposits had the role of being a major oxygen-sink. Only after such sinks had started to fill up – by consuming most of the available dissolved iron - did atmospheric oxygen begin to accumulate in any great quantity. That was not to happen until well into the Proterozoic.

A large block of banded iron formation, 2x3x1 metres in size and weighing about 8.5 tons.

WALES THE MISSING YEARS

Part seven: Plate Tectonics - the facilitator of biodiversity

Everest. To some, it's a thing that must be done, "because it's there", a quote attributed to the British mountaineer, George Mallory, in 1923. Today, the most popular route for summit-bids is the South-East Ridge. It starts at the wind blasted South Col, from where climbers set out from their high camp in the dead of night for the laborious slog ahead, all above the 8,000 metre contour. The air is rarefied at that height by some two thirds: no wonder oxygen cylinders are required by many. I've not attempted this route and am likely too old now, even if I wanted to have a go, but I have read enough to appreciate it involves strength and stamina of a kind not needed for a stroll up Plynlimon.

Slopes that vary in steepness from tiring to exhausting lead up and up to the South Summit, from where the knife-edged, heavily corniced ridge leads on via the precipitous Hillary Step to the top, a giddying drop of up to 3,000 metres on either side. But there are those climbers who also have geological tendencies, beginning with Noel Odell (1890-1987). Back in the 1920s, on another attempted route to the summit, he realised there's quite a lot to see when you look down, not at the drop but at the ground beneath your feet in those spots where bedrock protrudes from the snow.

Everest's entire summit-pyramid consists of marine sedimentary rocks - fossiliferous shallow water limestones - of mid-Ordovician age, around 465 million years old. What on Earth are they doing up there, the remains of an ancient seafloor ecosystem, now at the highest point on the entire planet?

Another question: one colossal difference between oceanic crust and continental crust is that while the latter may preserve geological features originating as far back as Archean times, most of the deep ocean floor is less than two hundred million years old. Why?

The answers to both lie in another critically important phenomenon on Planet Earth, unique among the rocky planets of our inner Solar System: plate tectonics.

Earth's brittle lithosphere is divided up into a number of slow moving tectonic plates, large and small. Their various movements relative to one another are familiar to us, since newsworthy events such as volcanic eruptions, earthquakes and tsunamis are focused along plate-boundaries. So: how did we find out about plate tectonics?

Plate tectonics was a controversial hypothesis when it was first put forward, to put it mildly, but is now accepted by an overwhelming majority of Earth Scientists. The roots of the hypothesis go back centuries and are of a geographical nature. Why, when looking at a globe, do Africa and South America look as though they ought to fit together? It is because they did once fit together, until about 110 million years ago.

In combination, Africa and South America originally made up part of the great continent, Gondwana. But 110 million years ago, they began to separate, the two landmasses drifting slowly apart from one another, towards their present positions. The ever-widening rift in between them was the juvenile South Atlantic Ocean, floored by newly-formed oceanic crust that had solidified from magmas continually erupting along the mid-Atlantic Ridge.

The existence of the Mid-Atlantic Ridge was known about as long ago as 1853 and confirmed in 1872 when traverses were made across it, making depth-soundings for the purpose of laying submarine telegraph cables. However, its geological significance was only to be realised decades later. Through this time, other mid-oceanic ridges were gradually discovered and mapped until it was realised that such ridges are a feature of all oceans. Typically rising 2,000-3,000 metres above the abyssal plain, they constitute the longest mountain ranges on Earth, even though their crests are mostly deep underwater. Crests are marked by rifts, along which magma, coming up from the mantle below, is erupted.

We know all of that now, but it may come as a surprise to learn that the apparent fit between Africa and South America was first noticed as long ago as the 1590s, by Dutch map maker Abraham Ortelius (1527-1598). With incredible insight, he commented how the Americas must have been "torn away from Europe and Africa". To elaborate, Ortelius noted: "The vestiges of the rupture reveal themselves, if someone brings forward a map of the world and considers carefully the coasts of the three." That's a remarkable observation for the time.

With improving global maps over the following centuries, the observations of Ortelius became even more apparent. On top of that, the trickle of geological data, from both sides of the ocean, became a flood. By the early 20th Century, one of the historical giants of science, Alfred Wegener (1880-1930), first described his hypothesis of continental drift. The ideas were subsequently expanded upon in his book, entitled 'The Origin of Continents and Oceans', the first edition of which was published in 1915. You can buy the fourth edition online today. I'd like to think Wegener would have been pleased by that.

Wegener envisioned Earth's current configuration of continents as resulting from the breakup of an older, larger landmass. The continents would sometimes come together and at other times rift apart and move away from one another. He didn't just have the apparent fit between Africa and South America: he had other geological evidence aplenty. Rocks on the facing coastlines of those two continents matched one another in so many details: for example, the same fossils occurred in sedimentary rocks in parts of Western Africa and eastern South America, now separated by thousands of miles of ocean. How else could things like the same species of land-dwelling or freshwater reptiles have lived on the different continents, at the same time, all those millions of years ago, unless such landmasses were at the time joined together?

Although Wegener could not explain the physical mechanism for continental drift, evidence supporting his hypothesis continued to build in the following decades. To many geologists, though - who were perhaps less acquainted with the results pouring in from Africa, South America and elsewhere - the idea that continents were constantly on the move still seemed preposterous. We scientists tend to be a conservative bunch; we demand that extraordinary claims require extraordinary evidence. But some did accept the continental drift concept, such as famous British geologist Arthur Holmes (1890-1965), author of the equally famed 'Principles of Physical Geology'.

Magnetic attraction

The tide truly turned firmly in favour of Wegener's hypothesis from the late 1950s onwards. Much of the robust new evidence in support of continental drift came not so much from the continents but from out on the oceans. New techniques in geophysics and in offshore marine geology brought about this change, an important contributor being the palaeomagnetic record. Palaeomagnetism is the study of the properties, preserved in the rocks, of Earth's magnetic field in the past. So how do rocks go about preserving ancient magnetism?

Certain minerals, such as the common iron oxide magnetite, are naturally magnetic, as the name might suggest. Magnetite is almost always present as a minor component of igneous rocks like basalt. Although magnetite crystals are swept this way and that by the currents within a basalt lava flow, once the solidified magma cools to below 570°C a change takes place. Beneath that critical temperature, permanent magnetisation is retained, meaning those magnetite crystals record the strength and orientation of Earth's magnetic field at that point in time.

Octahedral crystals of magnetite up to 2 mm in size from an iron ore deposit in North Wales. This rich sample makes the needle of a compass deviate from today's magnetic north by up to 60°.

Such ancient magnetism is the reason why a compass can be unreliable for navigation in mountains where magnetite-bearing igneous rocks are common, as climbing guidebooks to such places will warn. The presence of all that magnetite can affect the compass needle, making it deviate significantly from today's magnetic north. That's a serious hazard in pathless terrain if the visibility is poor.

Palaeomagnetism also records how Earth's magnetic field has frequently switched polarity through geological time, so that north has become south and vice-versa. Such switches are referred to as 'magnetic reversals', with the magnetic record of the past featuring irregular but frequently alternating periods of normal and reversed magnetism. The process is not cyclic as such: there have been several such switches in the last few million years but in the deeper past, the magnetic field has periodically settled into one or the other states for much longer.

When palaeomagnetic data are plotted on a column of geological time, the convention is to shade periods with normal magnetism, like now, in black. Periods of reversed magnetism are left in white. The result, in the finished plot, is a repeated alternation between black and white through time, looking a bit like the stripes on a barcode.

Critically, palaeomagnetism in rocks can be measured. Technology developed in the Second World War for submarine detection, consisting of sensitive magnetometers that could be towed behind boats, made such measurements possible. Specialised ships were deployed, to undertake repeated traverses over the mid-oceanic ridges and measure the strength and polarity of the palaeomagnetism preserved in the basalts of the ocean floor. Some boats were equipped to drill down into these rocks and retrieve samples from them, making it possible to determine their mineralogy, chemical composition and age. Long cruises and multiple boreholes thus made it possible to build up a picture of the ocean floor in unprecedented detail. Once the palaeomagnetic data were plotted on maps, the results were spectacular. In either direction away from the mid-ocean ridge, there were those striped barcodes of normal and reversed magnetic polarity, like a mirror image of each another, recorded by the rocks.

Magnetic striping - as the stripy pattern was called - showed how the ocean floor was made up of rocks that originated from the continual eruptions of basaltic magma along the mid-ocean ridge. The magmas solidified to form ocean floor basalts, locking into them the status of Earth's magnetic field at the time. As new magma was constantly being erupted at the mid-ocean ridge, the older ocean floor, with its recorded magnetism, was continually pushed away from the ridge-crest in both directions. Thus was born the concept of seafloor spreading.

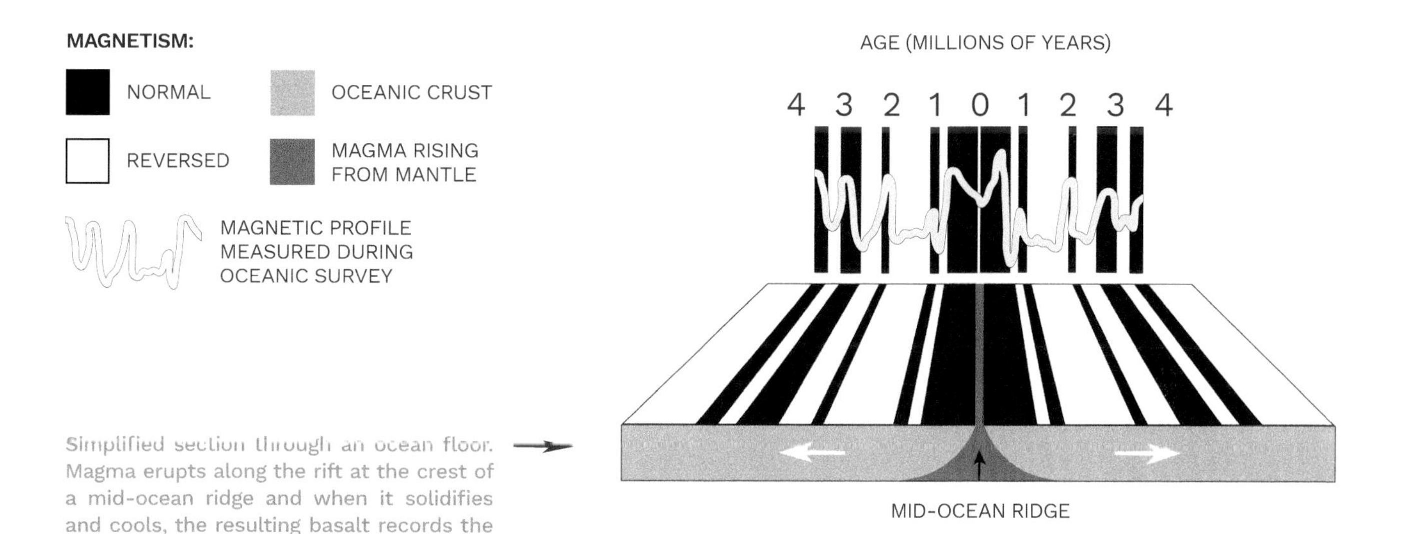

Simplified section through an ocean floor. Magma erupts along the rift at the crest of a mid-ocean ridge and when it solidifies and cools, the resulting basalt records the orientation and strength of Earth's magnetic field at the time, a property measurable using sensitive magnetometers.

The second part of the equation, concerning the ultimate fate of oceanic crust, was solved using seismic imaging, a geophysical technique that plots the varying speed of shock waves as they travel from a natural or artificial source through different kinds of rocks. Such 'seismic profiles' showed that in the deep ocean trenches, mostly adjacent to the margins of continents, oceanic lithosphere was being forced down into the depths of the mantle, consumed by the process termed 'subduction'.

Plate movement is imperceptibly slow: the fastest-moving plates are cruising along at a similar speed to that at which human hair grows. Most of the plates are moving at less than the rate at which human fingernails grow. However, when you consider the colossal volumes of rock involved, that slowness becomes somewhat irrelevant. Two continental plates, colliding head on at even that slight a speed, can carry enough force to seriously crumple the rocks in the collision-zone, relentlessly pushing them up into a Himalayan-sized mountain chain over several million years. The Himalayas were formed in this way, as India drifted northwards to plough into the rest of Asia.

Apart from continental collision-zones, the boundaries between Earth's tectonic plates are divided into four key types, defined by how the plates are moving relative to one another. Firstly, along mid-oceanic ridges, continuous eruptions of basalt generate oceanic crust, forming plates that are gradually moving away from one another. Mid-oceanic ridges are therefore known as 'divergent' or 'constructive' plate-boundaries. As the crust slowly moves away from the ridge in either direction, it and the uppermost mantle beneath it - the asthenosphere - are cooled by the conduction of heat into the overlying cold abyssal waters. That conduction and cooling gradually converts ductile asthenosphere into brittle oceanic lithosphere. As a consequence, the older the oceanic crust, the more brittle oceanic lithosphere is sat underneath it, because it has had longer to cool in this manner.

Different rates of seafloor spreading occur along different sectors of a mid-oceanic ridge. This is due to variations in the rates at which new oceanic crust is being produced from one sector to another. Such differences set up strong physical stresses within the oceanic plates. Being brittle, the plates eventually fracture along fault lines. Movements along such faults, causing underwater earthquakes, temporarily and repeatedly relieve that stress.

Active plate boundaries showing the plate movements:

DIVERGENT OR CONSTRUCTIVE PLATE BOUNDARY

CONVERGENT OR DESTRUCTIVE PLATE BOUNDARY

TRANSFORM PLATE BOUNDARY

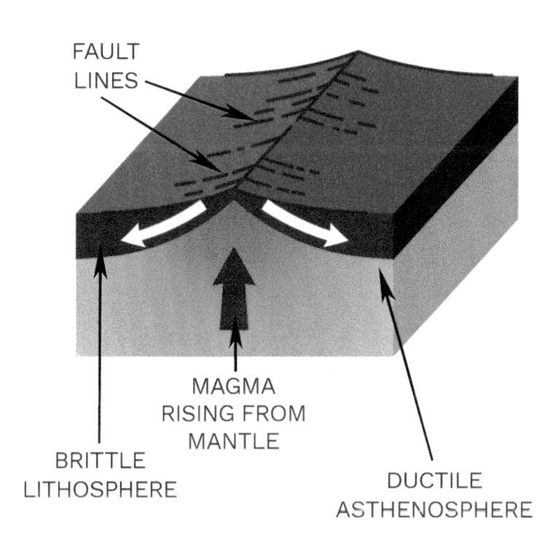

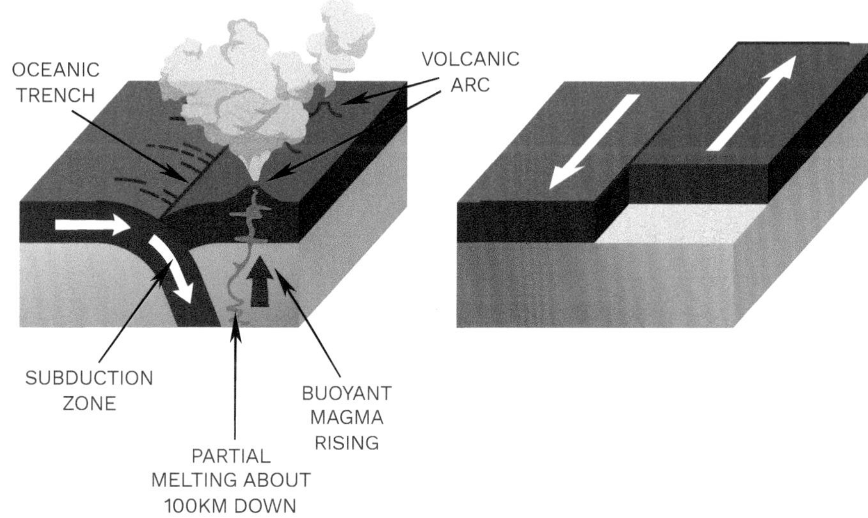

Subduction - how oceanic crust meets its fate

While there is room for ocean expansion, nothing much happens along the boundaries between oceanic plates and the adjacent continents. Geologists therefore refer to such inactive plate boundaries as 'passive'. The boundary between the continental crust of Europe and the oceanic crust of the Atlantic is a good modern example. But eventually, the expansion of the Atlantic will cease. The continents on either side of the ocean may stop moving apart or even begin to move back towards one another. Oceanic crust will then begin to be consumed at subduction-zones, which are known as 'convergent' or 'destructive' plate boundaries. Such boundaries may involve an oceanic and a continental plate or, in some cases, two oceanic plates.

Within an active subduction-zone, oceanic crust, together with the lithosphere beneath it, is driven relentlessly beneath the leading edge of the overriding plate, diving down into the depths of the mantle. Try to imagine that for a moment: the sheer volume of rock involved and the tectonic brute force applied to it. These are forces almost beyond human imagination. As the plate descends into the bowels of the Earth, it encounters increasingly hostile physical conditions, with gradually increasing temperatures but, in particular, phenomenal pressures.

Under the extreme physical conditions encountered in subduction-zones, metamorphism of the oceanic lithosphere is guaranteed and the rocks produced are distinctive, recording the moderately high temperatures but extremely high pressures. At depths of around a hundred kilometres or so, ordinary, dull grey basalt, already having been modified considerably, completely recrystallises into the bright green and red spotted rock known as 'eclogite'. As rocks go, it's spectacular stuff.

Relatively uncommon at Earth's surface, since the geological processes required to transfer rocks from the upper mantle to the upper crust are a big ask, there's plenty of eclogite down in the mantle because of all the old 'fossil' subduction-zones that are dotted about the planet. The physical mechanism for the process by which such rocks can be returned to the surface, known as 'exhumation', is still not fully understood - but evidently such things do happen. For this reason, eclogites are a major area of geological research.

Eclogite – a rock formed at extreme pressure. Sample (7cm across) from Norway, with crystals of red garnet and bright green omphacite, the latter's name derived from the Greek omphax or unripe grape!

Long before the basalt-eclogite transformation depth is reached, the increasing heat and pressure have other important effects, a key one of which is dehydration. Old seafloor rocks that have had an ocean sat on top of them, for as long as 200 million years in some cases, contains lots of hydrous, or water-bearing minerals. Such minerals may lead to water making up several percent of the chemical composition of subducting oceanic plates.

Some hydrous minerals let go of that water easily, others more stubbornly, requiring higher temperatures to drive it off. That in turn means dehydration is a progressive process, occurring all the way down into and in some cases beyond the eclogite depth of formation, under increasingly harsh circumstances. The amount of water released to the mantle, through dehydration-reactions during subduction, is colossal.

That water plays a vital role because firstly under the prevailing physical conditions, it actually lubricates the movement of the plates as they are forced past one another. Secondly, the melting point of rocks is significantly lowered in the presence of water: it acts as a flux. At the temperatures prevailing many tens of kilometres beneath the surface, the water helps parts of the descending oceanic plate and the adjacent mantle to start to melt.

Partial melting of rocks in the depths of subduction-zones is a widespread process, important both in the past and on today's Earth. It leads to the production of large volumes of magma that, being buoyant, make their way upwards through any structural weaknesses, such as fractures, in the overriding plate. Some of the magma will make it all the way to the surface, to be erupted from volcanoes, whereas a lot will solidify beneath the surface to form intrusions. Either way, the recycling of oceanic lithosphere by subduction, followed by magma generation then volcanic and intrusive activity, is how new continental crust is formed.

Commonly, such subduction-related volcanoes form arc-shaped chains of islands, rising up through the relatively shallow seas of a continental shelf or the deeper waters above an overriding oceanic plate. The Pacific Ocean's "Ring of Fire" provides excellent examples of such volcanic arcs. Unlike the exclusively basalt-spewing mid-oceanic ridge volcanoes, such 'island arcs' erupt a wider range of magmas, in terms of their chemical composition. That variability in chemistry can occur for two key reasons.

Firstly, hot magma rising up through the crust can become chemically contaminated as it interacts with the varied rocks it encounters. Some crustal rocks may undergo partial melting themselves, the new melt joining the original magma in that upward journey. Secondly, modification can occur if a body of magma collects for a protracted time in a magma-chamber deep underground. As the magma slowly cools, some minerals start to crystallise out from the liquid, those with the highest melting points crystallising first. Such crystals will sink down through the magma and accumulate on the chamber floor. Our old friend, the iron and magnesium silicate olivine, is a classic example: a common rock-forming mineral, it has a very high melting point so will readily crystallise in such a manner. As it does so and the crystals sink and gather on the chamber floor, the remnant magma becomes progressively iron and magnesium-impoverished. Such progressive changes will be reflected by the chemistry of the resulting eruptive or intrusive igneous rocks.

Intrusive rocks, of island arc affinity, from the early Ordovician of North Wales. Older, iron-rich magma has cooled and crystallised into gabbro (dark greenish-grey). The gabbro has then been invaded by a younger, iron-depleted and volatile-rich magma that produced the speckled, whitish rock, microdiorite. →

The final type of plate-boundary, the 'transform' type, is where two plates move side-by-side relative to one another. In such circumstances, neither is new crust formed nor old crust consumed. However, the movements are not bereft of consequences. Cold, brittle lithosphere does not move smoothly. Ever tried cutting firewood with an old, rusty and blunt bow-saw? It's an unpleasant experience: the teeth cut in a bit, stall and then cut in again in a series of jerks. A similar thing happens with tectonic plates moving side-by-side: they stick for a bit, then in a sudden jerk they move and the physical stress is relieved, before it starts to build up once again. Each jerk produces an earthquake. The major San Andreas Fault in California is a modern day plate boundary of this type and as the locals know only too well, the quakes can be phenomenal in their power.

What causes Earth's tectonic plates to move? That is still a subject of debate among geologists with more than one competing hypothesis. Movement is likely facilitated by the ductility of the asthenosphere underlying the plates, but that doesn't explain what drives the motion. One school of thought, first proposed by Arthur Holmes as far back as 1928, cited Earth's internal heat causing convection currents in the mantle, the vertical motions inflicting stresses and thereby movements in the overlying plate. Another hypothesis is that there are two major controls on plate movement, both involving oceanic crust. One is the pushing effect exerted by newly-formed oceanic crust at mid-ocean ridges, against older seafloor to either side. This is known appropriately enough as 'ridge-push'. The other is the pulling effect of oceanic crust diving down into the mantle along subduction-zones, known as 'slab-pull'.

A descending cold oceanic plate is denser than the hot mantle it is forced into, so that once subduction is underway, the plate will continue to dive downwards under its own weight. Slab-pull can be so strong that if subduction ceases for any reason, the part of the plate that has already descended can break off completely.

Once plate tectonics was ongoing on Earth, a constant cycle was set in place, involving oceans opening and widening at constructive plate boundaries and closing at destructive plate boundaries, as the continents slowly drifted around the globe. At times, continents collided with one another to form bigger landmasses, or rifted apart to form smaller ones. On a few rare periods along the geological timeline, the continents were all joined together, to form a supercontinent – Pangaea is the famous one, but there have been others, further back in the deep past.

Making mountains

Collisions between continents can occur at any angle, from head on smashes to glancing blows. As a consequence, they vary enormously in their effects. The most violent examples can lead to dramatic deformation of rocks, as they are folded, variably metamorphosed and thrust up and over one another in great sheets. That is the reason why there are marine fossils in sedimentary rocks, originally deposited on a seabed, close to the summit of Everest: the rocks got shoved up there. Such mountain-building events have a technical name: 'orogeny', a word stemming from the Greek: oros (mountain) and genesis (creation). The study of rocks that have been deformed during orogenesis is a specialised area of science known as structural geology, with its practitioners able to unravel the complex chains of events leading to the formation of the world's great mountain-belts.

Orogenesis has many side effects, one of the most important in economic terms today being that it brought about the global slate industry. Slate is formed when fine-grained sedimentary rocks like mudstones or siltstones are strongly compressed by tectonic brute force. Such rocks contain a great abundance of minerals known as 'sheet-silicates', these forming tiny flakes that split apart very easily. In an undeformed sediment, such flakes may be arranged in a higgledy-piggledy fashion. But in response to compressive force, all the flakes recrystallise at right angles to the direction from which that compression is coming, so they become arranged parallel to one another. The resultant rock can be split into thin sheets along that alignment, known as the rock's cleavage. We have plate tectonics to thank for that.

Descriptions of the effects of plate tectonics are to be found in modern accounts of the geological evolution of every part of the world. However, one of the most controversial questions in geology is as follows: just when did plate tectonics, as we know it today, begin? There are a number of competing hypotheses regarding this area of uncertainty. Scientific uncertainty doesn't mean we know nothing - instead it means we need to look at more aspects of a problem until the data can tell us what's going on - just as we have done with the current pandemic. It is scientific uncertainty that drives the continuous process of scientific research and always will as our understanding of our surroundings grows and grows.

Evidence for seafloor spreading, subduction, island arc volcanism and related processes is present in great abundance in the world's younger rocks, especially those formed during the past 750 million years or so. In older rocks, that's not always the case, since the evidence for continuous plate tectonics is more ambiguous. Some geologists contend that plate tectonics is a relatively recent development, starting perhaps a billion years ago. However, an alternative and reasonable view is that many of the oldest rocks are so difficult to study, due to severe metamorphism, that we're only just starting to unravel their well-guarded secrets. In places, such rocks are producing evidence that some form of plate tectonics was operating a lot longer ago, during the Archean, but whether the process was continuous or intermittent is not known. There could have been many false starts to the situation we see on the modern Earth.

What initiated plate tectonics in the first place is another area of controversy. Originally, the Hadean Earth was a seething ocean of magma, but as it cooled, a shell-like crust of solid rock would have formed at the surface and grown in thickness through continued, widespread volcanism. It is also important to reiterate that the early Archean Earth had differences to the familiar planet of today. Its mantle was several hundred degrees hotter. Its lithosphere was much thinner and weaker. As a consequence, the response of the lithosphere to forces acting upon it may have been different, compared to the situation we see in more recent geological time. But a switch from a solid shell to continuous, multiple tectonic plate-movements certainly happened at some point.

The realisation that oceanic plates are generated by continuous eruptions at mid-ocean ridges, are pushed progressively away from them and are eventually forced back down into the mantle at subduction-zones, was a defining moment in Earth Science. It utterly changed the way in which we look at the planet and its evolution. No wonder Earth's surface is not a mass of impact craters like that of the Moon: although impacts have of course occurred on Earth and some of the craters are famous, the planet's surface has been recycled time and again. Plate tectonics makes Earth a geologically dynamic, vibrant place, never resting, whereas the tectonically lifeless Moon still preserves an abundance of features created in ancient time. Indeed, so far as we know, Earth is the only rocky body in the Solar System upon which continuous plate tectonics is operative. The other rocky planets are, by comparison, dead worlds.

On Earth, plate tectonics creates a vast range of life-supporting habitats, from the highest mountains to the depths of the ocean trenches. Can you imagine a world without plate tectonics? Only a fraction of those diverse habitats would be available.

That constantly ongoing cycle of subduction, dehydration, melting and volcanism also recycles carbon dioxide, water and other volatiles in and out of Earth's mantle. Carbonate minerals, common in ocean floor rocks, go down with subducting oceanic plates and carbon dioxide is liberated when they are cooked during the metamorphism and partial melting. Incorporated into the resulting magmas, the gas is then transported back up to Earth's surface and released from volcanoes. As a consequence, plate tectonics has been of critical importance in regulating Earth's greenhouse effect over geological time - before we came along and started tinkering with it ourselves.

If plate tectonics completely stopped, which may happen in the geologically distant future, widespread volcanism and mountain-building would be no more. Existing mountain ranges, like the Alps and Himalayas, would bit by bit be worn down by erosion and sediment removal. The end result, many millions of years on, would be subdued landscapes of low rolling hills and extensive plains stretching into the far distance. Loss of the mountain ranges would have profound effects on regional climates. Not only that, in the absence of volcanic arcs spewing out carbon dioxide, Earth's greenhouse effect would diminish in strength. In a nutshell, the planet would become a cold, miserable place complete with uninspiring scenery. We have a lot to thank plate tectonics for, despite the natural disasters it can cause.

No matter when or how plate tectonics commenced, there still remained other bits of the equation that needed to come into place in order for advanced multicellular life to be able to thrive on Earth. Probably the most important of these is something we use, without a thought, every waking or sleeping moment: breathable air.

WALES THE MISSING YEARS

Part eight: The Great Oxygenation Event, the Ozone Layer and Earth's Carbon Cycle: what could possibly go wrong?

Many years ago, I spent two weeks on a high-level tour of the Stubai and Otztal Alps that straddle the Austria-Italy border. It's a great way to see the mountains, staying at high mountain huts and leaving each day before dawn for the climb ahead. Sunrise on a glacier, the frozen snow sparkling with frost and the Dolomites silhouetted in the distance, were all regular, magical features of that trip. You start early so you can climb the day's peak then get off the glaciers before the snow gets too soft due to the afternoon heat, thereby making hidden crevasses more hazardous. Then it's on to the next hut for food, beer and a sunbathe. I had trained hard for the trip by running several miles a night and having tough walking weekends in Wales. Nevertheless, day one was a shock to my system.

On that first day's climb, the objective was a twin peak by the name of Feuerstein, 3,268 metres high, or 10,720 feet in old money. The hut where we stayed the night before was at around 2,500 metres, so I was not expecting a difficult morning. With it being my first day at altitude, though, it was exhausting. Even at that height the air is rarefied - the atmosphere is densest at sea level, with an overhead pressure of about 1013 millibars on average. The higher you go, the less the atmospheric pressure - on Feuerstein's summit we're looking at around 680 millibars - and therefore less oxygen is available - around a third less in this case.

The human body is taken by surprise when it first encounters rarefied air. It initially responds to the ambush in a primitive fashion, by faster breathing and a big increase in the heart rate, both of which make you feel lousy. After a day or two at height, things normally improve: your body has made the necessary metabolic adjustments in order to cope with the new regime and the climbs become pleasant as opposed to purgatory. You have become acclimatised to the altitude.

Here's something to ponder: at the start of the Proterozoic Eon, 2,500 million years ago, you wouldn't have been able to breathe at all. There was virtually no oxygen present in the air: nothing to acclimatise to.

Out of all the things that happened to Earth during the Proterozoic Eon, the most important was probably the Great Oxygenation Event, something that would change the planet forever. Taking place progressively from around 2,450 million years ago, this was a point of no return for Earth's atmosphere. In its wake, the rate of supply of oxygen to the air, via the photosynthetic activity of primitive but abundant microbial plant life such as the cyanobacteria, outstripped the rate of consumption of the gas in sinks. For the first time in Earth's history, the atmosphere was oxygenated on a permanent basis.

It was tiny cyanobacteria like this modern example that changed the world forever. The filaments are up to tens of microns in length and a few microns in thickness, so you need a microscope to study them.

Image: Willem van Aken, CSIRO, Australia.

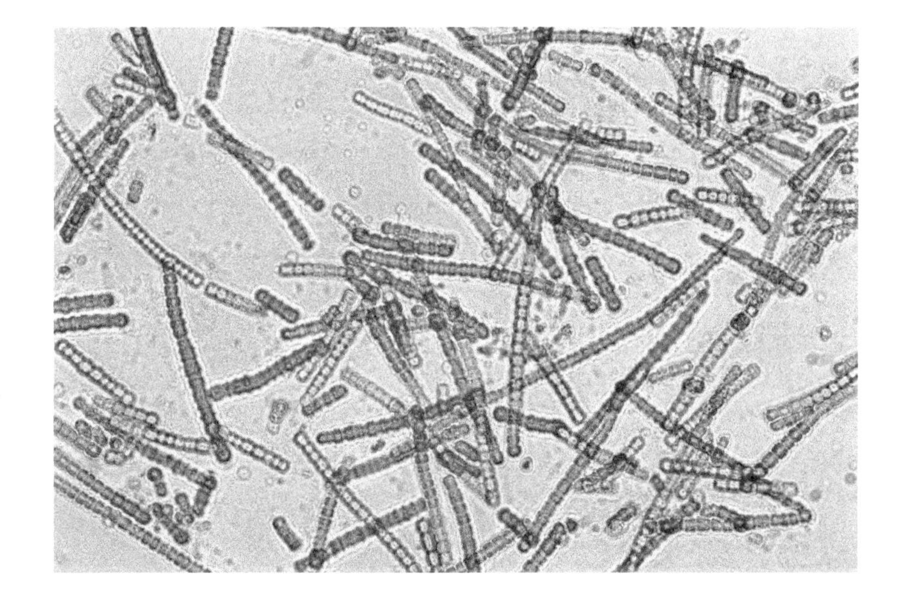

The first 'red-beds' appear in the geological record at this very time. Red-beds are sedimentary rocks, like sandstones or mudstones, deposited on land under oxidising conditions: their reddish colour is due to the abundant presence of fine-grained red iron oxides such as hematite. Their presence is direct evidence for an oxygenated atmosphere. Prior to the Great Oxygenation Event, there were no red-beds: iron, released by chemical weathering of rocks, remained in solution and was transported into the oceans, to form those banded iron formations on the deep seafloor. After the atmosphere became well-oxygenated, banded iron formations became a relatively scarce feature in the geological record: most of the iron instead remained on land.

← Red beds of Triassic age, only just over 200 million years old, forming the lower part of a sea cliff at Penarth in South Wales. Such distinctively-coloured rocks, deposited in a terrestrial and oxidising environment, first appeared in the geological record in the Proterozoic Eon, following the oxygenation of Earth's atmosphere. The white bands are evaporites, consisting mostly of gypsum, representing the remnants of dried up seasonal salt lakes.

Fundamental planet-wide changes, such as oxygenation of Earth's atmosphere, often bring with them both benefits and pitfalls. For example, the oxygenation was without doubt an absolute disaster for anaerobic life, vastly reducing the number of available ecological niches for these oxygen-intolerant critters. Another name for the event is the Oxygenation Catastrophe. It was, so far as we know, the first instance of one class of living things changing the planetary environment so drastically that the very existence of another group of organisms was badly compromised.

In terms of benefits, once there was a significant amount of free oxygen in the atmosphere, the stratospheric Ozone Layer started to form – a definite plus. Ozone (O_3) is formed by the bombardment of oxygen molecules (O_2) by incoming solar radiation. The bombardment splits the O_2 molecule into two oxygen atoms, which are then able to combine with other oxygen molecules to make the O_3. Ozone is extremely effective at absorbing ultraviolet rays, including, vitally, the harmful short-wave ionising radiation referred to as UV-B and UV-C. The gas therefore acts as a shield, but this defensive feature of our planet took a considerable amount of time to reach its full strength; it had likely become effective by the late Proterozoic. In turn, protection from harmful wavelengths of UV would make Earth's land surfaces prospectively habitable for the first time.

In the evolving early Proterozoic atmosphere, the progressive increase in oxygen content affected the amount of methane present in the air. Atmospheric methane reacts readily with oxygen, with carbon dioxide and water vapour the end products of the process. These products are still greenhouse gases but compared to methane they are much less potent. Furthermore, the continued spread of photosynthetic life, with its hearty appetite for carbon dioxide, consumed a large amount of that gas. An overall reduction of the greenhouse effect, due to falls in the concentrations of its key perpetrators, might well be expected, and that is exactly what happened. Temperatures fell sharply, plunging the world into a large-scale ice age.

The advances and retreats of the ice sheets, collectively known as the Huronian glaciation, are recorded in rocks of that age all around the planet. The overall cold period lasted for over two hundred million years. To understand what drives such dramatic and potentially catastrophic climate perturbations, we need now to examine the key driver of such things: Earth's carbon cycle.

Earth's vitally important carbon cycles, fast and slow

Fluctuations in Earth's carbon cycle are the prime regulating mechanism of the greenhouse effect: the cycle therefore acts as a kind of planetary thermostat, a control knob that regulates global temperatures over prolonged time periods. Chuck a wrench into that machine and you get problems - big problems. So let's see how it works and how it can go wrong.

Earth's carbon cycle has two key components, a fast part and a slow part. The fast carbon cycle involves the movement of carbon through the biosphere over short time spans. Key participants are photosynthetic plants, which use the energy received from sunlight to react carbon dioxide with water within their cells, to form sugar and oxygen. Sugar reserves are stored away in stems, tubers or roots, depending on the species. During the growing season, plants need to tap into those reserves, breaking down the sugar. Many animals eat plants and also break down that sugar, in order to obtain energy. When plants die they decay, their foliage being consumed by bacteria, fungi, or sometimes fire. In all cases, the chemical reactions that are involved have the same end products, carbon dioxide and water vapour, both of which end up back in the atmosphere.

Due to these chemical reactions, a noticeable feature of the fast carbon cycle is that it causes atmospheric carbon dioxide levels to fluctuate in a regular, seasonal pattern, like a heartbeat. It is the echo of the Northern Hemisphere's growing season, because that's where more of Earth's land surface is situated, so there's more photosynthetic plant life. In the Northern Hemisphere winter, when many plants are either dead or dormant, atmospheric carbon dioxide levels rise, the reverse happening in the spring and early summer when the growing season is at its height.

In this way, a kind of equilibrium is preserved, with seasonal peaks and troughs that cancel each other out. Well, that would be the case if we were not going through Earth's fossil fuel reserves as if tomorrow did not matter. The seasonal changes in carbon dioxide levels instead form regular, symmetric wobbles on the upward slope that represents our addition of carbon dioxide to the atmosphere.

Mauna Loa observations, 1999 to mid-2019. Data: NOAA

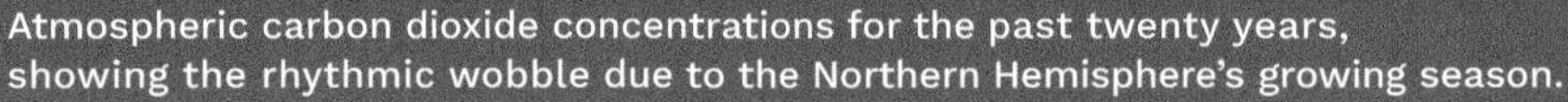

Atmospheric carbon dioxide concentrations for the past twenty years, showing the rhythmic wobble due to the Northern Hemisphere's growing season.

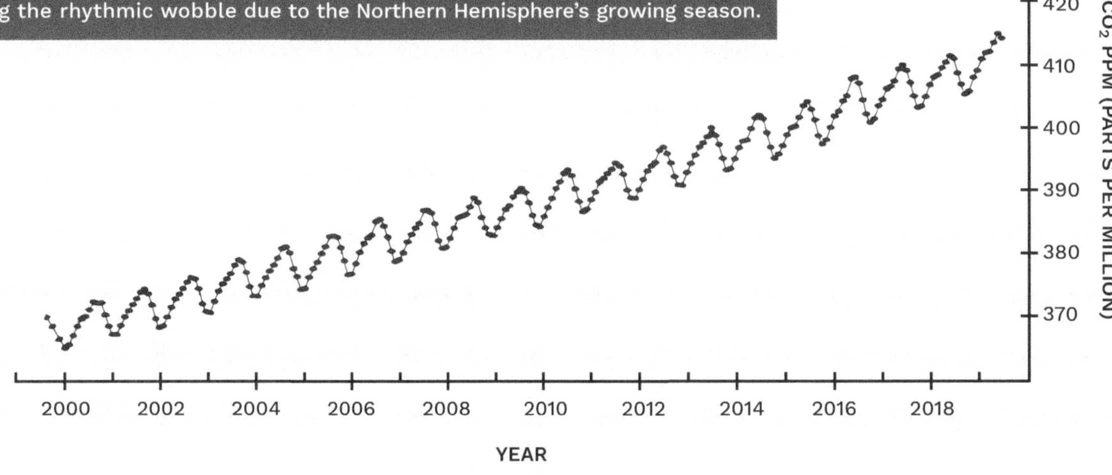

The fast carbon cycle has an annual heartbeat on top of the steady rise in carbon dioxide caused by human activity.

In stark contrast to the fast carbon cycle, the slow version operates over geological timescales and has affected atmospheric CO_2 levels, and therefore temperatures, a great deal more over Earth's lifetime. The reason why the slow carbon cycle is so important is because many of the geological processes that lead to long-term changes in CO_2 levels - either upwards or downwards - take place over very long periods and do so on an erratic basis.

Annually, up to a hundred million tonnes of carbon pass through the slow carbon cycle, due to natural processes. That's small compared to the fast carbon cycle, through which up to a thousand times more carbon passes every year. However, remember that the fast carbon cycle is a give-and-take seasonal process, whereas the slow carbon cycle runs in one direction or another over periods often measured in millions of years.

Whilst considering such numbers, where do man-made CO_2 emissions fit in? When one considers the nature of the fossil fuels that we burn, we are clearly tapping into a carbon source that was trapped, for many millions of years, within part of the slow carbon cycle – until we deliberately dug or pumped it up and used it for energy. That represents an exceptional event along the geological timeline: no other species has done anything of the sort. Our activity of this nature currently releases some 36,000 million tons of CO_2 into our atmosphere every year.

Returning to the non-human related workings of the slow carbon cycle, we will firstly consider carbon sources, by far the most important of which is volcanic activity. Today, according to the United States Geological Survey, the world's volcanoes release between 180 and 440 million tonnes of carbon dioxide to the atmosphere every year - between 200 and 80 times less than our man-made emissions. However, there have been widely-spaced times in Earth history, tens of millions of years apart, when abnormally intense periods of volcanism have occurred, forming what are known as 'Large Igneous Provinces'. On occasion, such thankfully uncommon events have caused carbon emissions to approach or even exceed the man-made ones of today. Large Igneous Province eruptions are of global geological significance, so we'll look at a well-studied example in a bit.

In more normal circumstances, carbon emissions from volcanic activity chiefly involve arc-type volcanoes sat above subduction-zones, within which carbonate-rich oceanic crust is constantly being recycled. But such sources of emissions are broadly compensated for by the carbon that gets locked up in long-term geological reservoirs. One of the most important examples of such a storage facility is limestone, a very common sedimentary rock consisting mainly of calcium carbonate. Another is organic-rich deposits like coal or oil shale. The pathway taken by carbon dioxide when it leaves the atmosphere and ends up in reservoirs like limestone involves a key process: chemical weathering. Weathering is a broad term covering the ways in which dissolved, corrosive substances, in rain and groundwater, attack and chemically break down the minerals making up the rocks at or close to Earth's surface.

The key ingredient in weathering is carbon dioxide dissolved in rainwater. A solution of carbon dioxide in rainwater is weakly acidic: remember, the old name for carbon dioxide is 'carbonic acid gas'. Rainwater containing carbonic acid is able to react with many different minerals although the rate of this reaction is highly variable from one mineral to another.

Some naturally occurring minerals are extremely stable. Think about gold, eroded mechanically from ore deposits and then recovered by prospectors, maybe hundreds of thousands of years later, by panning river-gravels for flakes and nuggets. Or consider quartz, silicon dioxide, found as hard, white pebbles on beaches. They're both pretty bombproof.

Other minerals, especially metal sulphides like pyrite, are in stark contrast extremely vulnerable, reacting rapidly with any moisture. The problem with sulphides is that once the weathering reaction starts, sulphuric acid is generated, a much more aggressive reagent. Once that gets in on the act, it's game over, as anyone who keeps or curates a mineral collection will know: specimens thus affected will in due course fall to bits.

Most common minerals lie somewhere in between these two extremes. But over geological time spans, dilute carbonic acid can dissolve away huge quantities of minerals such as the silicates of sodium, calcium, magnesium and iron making up, for example, that abundant igneous rock, basalt.

Weathering is affected by temperature: the warmer the climate, the faster the process, because most chemical reactions proceed more rapidly at higher temperatures. On Earth today, the deepest and most intense weathering is to be found in the Tropics, where it can extend tens of metres down into the rock. The reason for that is because the climate is both very warm and very wet at times, so that the agents of weathering are delivered in large amounts and the high temperatures help them to get to work efficiently. Water-soluble elements like calcium, magnesium and sodium are dissolved, leaving behind insoluble minerals like quartz and converting other elements like iron into their insoluble oxides. This is why heavily weathered rocks tend to have a rusty appearance.

Weathering starts off superficially, giving the rock a bleached outer rind.

Given time, the weathering process can reduce solid rock to fragments and grit.

Image: Geomartin from Wikipedia, licensed under the Creative Commons Attribution-Share Alike 4.0 International license.

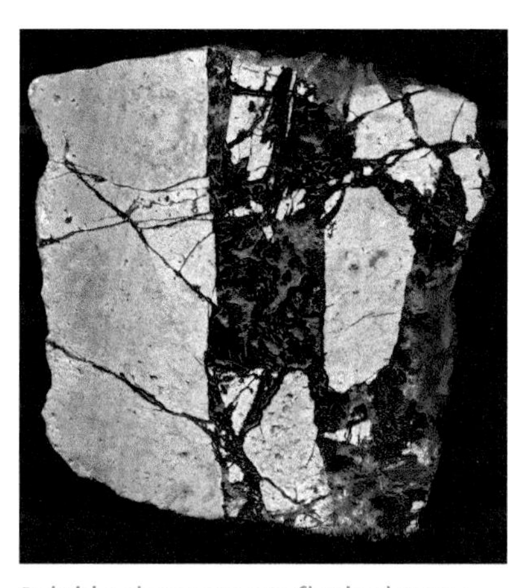

Soluble elements are flushed away in rainwater, leaving behind insoluble minerals like silica and this black manganese oxide. This bleached-looking rock's unweathered equivalent would be slate-grey in colour.

Why the sea is salty

Rainwater not only introduces carbonic acid on a continuous basis, but it also removes those dissolved weathering products, constantly flushing them out into the rivers and thence to the seas. Weathering of rocks thereby produces a variety of dissolved substances that end up in seawater. In any solution of any compound dissolved in water, the constituents of that compound will be represented by electrically charged atoms or molecules, known as ions. As an everyday example, table salt, which is sodium chloride, dissolves easily in water. The resulting solution contains ions of sodium and chlorine. If you let the water evaporate away, the ions will recombine and salt crystals will form.

Over geological time scales, various dissolved substances, liberated by the chemical weathering of rocks, have built up in the oceans. As well as the inflow of rivers bringing in the dissolved salts, the solutions have been further concentrated by evaporation of water at the sea surface. That's why, in a nutshell, seawater is salty. The oceans are enormous bodies of water, though and it is thought that it must have taken several hundred million years to get them to their current level of saltiness, which has probably remained pretty much stable ever since. This relative stability is because dissolved salts are also removed from the ocean waters by the chemical reactions that form various minerals in the sediments on the seafloor and in the underlying rocks. In that way, the system has reached a sort of equilibrium, where input and output pretty much balance each other out.

Carbon dioxide also enters the marine environment via two other key pathways. Firstly, there are the photosynthesising seaweeds and phytoplankton that inhabit the photic zone, that nutrient-rich, well-lit and highly populated upper layer of the water. In the clear waters of the open ocean, the photic zone may be as much as 200 metres deep; in muddy waters near estuary mouths it may be relatively shallow. Photic zone plants absorb carbon dioxide, convert it into carbohydrates and take those down to the seafloor with them once they die.

The second pathway for carbon dioxide to enter the marine environment involves its dissolution at the air-water interface, to form a weak solution of carbonic acid. A solution of carbonic acid consists of hydrogen ions and bicarbonate ions; under normal circumstances, the latter are easily converted into carbonate ions.

In the slow carbon cycle, the carbonate and calcium ions in seawater have a vital ecological role, because they are able to combine to form solid calcium carbonate. The calcium carbonate may either be deposited as an inorganic chemical precipitate or as a biogenic substance, produced by living organisms in the sea for various important purposes.

In today's oceans, as in those of the more recent geological past, biogenic calcium carbonate is produced from seawater by corals, in order to build their familiar skeletons that make up coral reefs. Bivalves, snails, sea urchins and many other common sea creatures, including tiny planktonic species, make their shells out of calcium carbonate. Plants like calcareous algae use it to reinforce their cells. The plankton in particular are important generators of carbonate, largely responsible for the calcareous 'ooze', a soft carbonate mud that thickly coats parts of the modern seafloor and is made up of planktonic remains that have sunk down to the seabed after the tiny critters have died. Together, these are all known as 'calcifying organisms'.

There are two different varieties of calcium carbonate, these being the minerals calcite and aragonite, chemically identical in terms of their composition but having different crystalline forms in terms of how their constituent atoms are arranged. Some calcifying organisms only make calcite when building their shells and other structures whereas others just use aragonite. In either form, the calcium carbonate tends to be chemically stable in shallow waters, since carbonate ions are present in such high concentrations that the water is said to be saturated with them. Such an abundance of available carbonate ions means that making calcium carbonate sea shells and other biogenic structures is a doddle.

Wherever calcium carbonate precipitation takes place from sea water, both inorganic and biogenic calcium carbonate accumulates on the seabed, forming often substantial deposits of sediment. Over geological time, such carbonate-rich sediments become buried and harden into limestone. Carbon locked up in limestone can be stored in that way for many tens if not hundreds of millions of years, which is precisely why that rock is one of the more important long-term carbon reservoirs on Earth. So far so good. But there's a catch. The chemical reactions, beginning with weathering and the dissolution of carbon dioxide in seawater and ending with the formation of solid calcium carbonate - are reversible – meaning that they can go both ways.

Fossilised calcifying organisms

Ammonites from the Lower Jurassic of the East Midlands.

Sea urchins from the Middle Jurassic of Cardigan Bay.

Corals from the Lower Carboniferous of north-east Wales.

A relatively recent bivalve from the late Pliocene of Suffolk - this one, the Glycymeris clam, is still with us today.

Dependence upon chemical reactions that can go into reverse is a potentially major issue to any marine organism whose very existence relies on being able to both make and maintain a calcium carbonate skeleton or shell. When conditions favour the reaction going one way, solid calcium carbonate is easily produced and preserved, but if they favour the reaction going in the other direction, that carbonate is dissolved. So what kind of conditions disrupt this part of the slow carbon cycle and cause things to go wrong?

Ocean acidification explained

Have you heard of the phenomenon known as 'ocean acidification'? It's in the news fairly often but not as much as climate change, although both are caused by the same thing – rapidly overloading the atmosphere with man-made carbon dioxide. Ocean acidification has the potential to cause major and widespread disruption of marine ecosystems, but what does it actually mean?

Acidity of water is measured by the pH scale, this being an inverse expression of the concentration of dissolved hydrogen ions: the greater the concentration, the stronger the acid and the lower the pH value. As a consequence, the pH scale runs from 0, which is highly acidic, to 14, which is highly alkaline. A pH value of 7 means that a solution is neither acidic nor alkaline: it is neutral. The pH scale is logarithmic, meaning that for each step downwards along the scale, towards zero, the increase in hydrogen ion concentration is tenfold. As a consequence, a solution with a pH of 3 has ten times more dissolved hydrogen ions than one with a pH of 4 – it is ten times more acidic.

Rainwater, with its dissolved carbon dioxide, typically has a slightly acidic pH of around 5.6. Seawater, being slightly alkaline, has a typical pH of around 8.1. Acidification simply means lowering the pH value from any point on the pH scale, in the direction of zero. If the pH of seawater shifts from 8.1 to 7.9, that is acidification, even though the pH is still on the alkaline side of neutral. To use an analogy, if the air temperature rises from -30°C to -15°C, it has warmed, even though it's still freezing cold.

If atmospheric carbon dioxide levels rise rapidly then the oceans will absorb more of the gas, just as is happening at present. That means the oceans will contain more carbonic acid, so what you end up with is seawater with increased amounts of both hydrogen and bicarbonate ions. Critically, with an increase in the concentration of hydrogen ions in seawater, carbonate ions start to abandon their calcium carbonate partnership: they slope off instead with the hydrogen to form more bicarbonate ions. As a consequence, if seawater pH falls, even by a small amount, the concentration of carbonate ions decreases significantly. Depletion in carbonate ions makes it much harder or in the worst cases impossible for shellfish and other calcifying organisms to build and maintain their protective shells and other structures.

Deeper, dark and cold abyssal ocean waters are, in contrast to the sunlit shallows, an unfriendly place for calcifying organisms. That's because such waters are almost always undersaturated in carbonate ions. This literally means those ions are far less available. Carbonate is a lot more soluble in seawater under the low temperatures and high pressures found in the depths. It prefers to stay in solution. Under such conditions, solid calcium carbonate will tend to dissolve.

The depth beneath which undersaturated waters are encountered, the 'saturation horizon' as it's known, varies considerably in different oceans because of the complex patterns of ocean circulation. For calcite in today's oceans, the saturation horizon is typically about 500 to 4,500 metres deep with the deeper values being found in the Atlantic and the shallower ones in the Pacific. For the more soluble aragonite, the saturation horizon is less than 500 metres deep across much of the northern Pacific, increasing to around 2,000 metres in the North Atlantic.

Ocean acidification is a potentially massive problem because it forces the saturation horizon to rise – it becomes much shallower. When that happens, more and more calcifying organisms find themselves exposed to carbonate-undersaturated water. If the rate of calcium carbonate dissolution outstrips the rate of its deposition, which is the worst case scenario in carbonate-undersaturated seawater, mass-mortality of calcifying organisms can occur. For example, bivalves start to lose the ability to create and maintain their shells. Without those, they are helpless.

Any mass-mortality event affecting calcifying organisms is seriously bad news because many such creatures occupy positions that are low down in the marine foodchain. That chain is underpinned by the single-celled photosynthetic phytoplankton, which use sunlight and carbon dioxide to produce organic carbon. They, in turn, form the baseline of a complex web of food supply to the 'heterotrophs', these being a diverse range of organisms from the tiniest zooplankton all the way up to the biggest sharks and whales, all of which by definition obtain their food and thereby energy directly, by consuming organic matter - other living things.

Clearly, if the food chain gets badly interrupted, especially anywhere near its lower levels as would happen if mass-mortality affected the calcifying organisms, there will be an adverse knock-on effect upwards and with the potential to radiate out in many directions. It's like blasting out the first floor of a tower block: the foundations may still be intact but the block collapses anyway. That is why ocean acidification, especially if it happens rapidly, can have devastating effects on biodiversity.

Rapid and catastrophic ocean acidification events have certainly happened in the geological past. They can usually be shown to have been related to sudden and massive atmospheric carbon dioxide overloads caused by Large Igneous Province-type eruptions. In some such cases, calcifying organisms, especially those that make aragonite shells or skeletons, have almost vanished from the fossil record for a few million years.

In normal geological circumstances, the slow carbon cycle is reasonably stable, although perturbations can happen and have happened in the past. As well as Large Igneous Province events causing rapid increases in carbon dioxide levels, there have been unusually intense and widespread weathering episodes that have drawn down abnormal amounts of the gas from the atmosphere. Such phenomena, significantly increasing or reducing the strength of the greenhouse effect and causing intense episodes of global warming or cooling respectively, inevitably lead to major climate perturbations. We'll now tale a look at two classic, extreme examples of such events: firstly the worst case scenario of a Large Igneous Province eruption and then the Snowball Earth glaciations.

The Siberian Traps: off the scale

In my view, the prime example of a Large Igneous Province eruption has to be the one that formed the extensive Siberian Traps volcanic district of eastern Russia, at the end of the Permian Period, some 252 million years ago. In this case, we'll step out of the ancient world to look at a relatively recent example because the rocks are less mangled so the research is easier to interpret and this topic is so important. Of all the geological misfortunes to affect the planet since the Archean, this was one of the worst.

The trouble began when a gargantuan volume of magma started working its way up from the mantle, through the overlying continental crust towards the surface. All eruptions involve such magma movements, but what made this one special was firstly the vast amount of magma involved and secondly that the molten rock forced its way up through a deep sedimentary basin, containing substantial deposits of oil, coal and evaporites. Interaction of these deposits with the magma created a vicious chemical cocktail of gases that were carried on upwards in colossal volumes, to be released in the eruptions. The ensuing pollution caused the biggest mass extinction in the fossil record, killing off 90% of marine species and 75% of those on land.

Let's just place the Siberian Traps episode in context. It's hard to imagine, even though any volcanic eruption can be deadly. Mount Saint Helens, in Washington State USA, erupted on May 18th 1980: this catastrophic event, the worst in U.S. history, was spectacularly caught on film. The entire northern flank of the mountain collapsed in a gigantic landslide, letting the magma chamber beneath depressurise in an instant. There followed a tremendous blast, in which the amount of energy released was calculated to be in the ballpark region of a 24-megaton nuke. In total, this nine-hour eruption spewed out some 2.79 cubic kilometres of lava, ash, and gases.

Cranking things up a bit, the eruption of Krakatoa on August 26th-27th 1883 was colossal, with its largest explosion being heard over 3,000 kilometres away in Western Australia. There was an energy release likened to a 200-megaton nuke and the volcano ejected an estimated 21 cubic kilometres of eruptive products.

Siberian Traps? Three million cubic kilometres at least: individual lava flows were up to 3,000 km^3.

The Siberian traps are an outstanding example of anomalous, 'within plate' volcanism, occurring far from any tectonic plate-boundary: in contrast, most of Earth's volcanoes are situated along plate-boundaries whether constructive or destructive. A modern example of within-plate volcanism is the huge caldera at Yellowstone: the last super-eruption there, 640,000 years ago, produced an estimated 1000 cubic kilometres of lava and ashes. Often, such centres of volcanic activity reflect plate movements, with a chain of volcanoes getting older and older as one goes back along it - as seen at the within-plate chain of ocean floor-rooted volcanoes forming the Hawaiian islands. Such age relationships led to the concept, first hypothesised in the 1960s, of a tectonic plate travelling over a fixed 'hot-spot' in the mantle below. In turn, the presence of convective 'mantle-plumes', within which heat was transferred from the core-mantle boundary towards the base of the lithosphere, became popular as an explanation for hot-spots in the 1970s.

Evidence favouring the mantle-plumes hypothesis is far from unequivocal, though, reflecting once more the difficulty in examining the deep Earth. Just because a hypothesis is popular it is not necessarily correct. Copious research on the subject has yielded sometimes contradictory data and the plumes hypothesis is still falling short of becoming a theory. Greater sophistication in deep seismology, something that is steadily improving, may in time bring the answers we need. We need not worry further about what caused the Siberian Traps but instead consider the timing and effects of the eruptions. It is instructive to look at how we discovered the relationship between the Siberian Traps eruptions and the mass extinction, since it is a great example of how modern geology works, with its multidisciplinary approach to such major tasks.

Firstly, palaeontology shows us when the extinction event took place. The fossiliferous sedimentary rocks in the internationally agreed type section for the Permian-Triassic boundary, at Meishan, Zhejiang Province, China, have long been scrutinised to a forensic degree. Successive individual beds have been numbered, their sedimentary characteristics described and their micro- and macro-fossils catalogued layer upon layer. As a consequence, the precise point in the sequence where the extinction occurred is well-defined. Above that bed, the fossils become relatively scarce and markedly less diverse, something you would expect to see if life itself had almost vanished. Correlation is to be found in many other Permian-Triassic boundary sections worldwide. That tells us something seriously big happened.

Secondly, radiometric dating has come to our aid, too, once again involving our old friend, zircon. Zircons may be rare in basaltic rocks and the Siberian Traps eruptions mostly produced basalt, but there are layers of zircon-bearing volcanic ash in between some of the basalt flows. Not only that, in the Chinese sections there are also layers of fine ash, between the sedimentary beds, also containing zircons.

Uranium-lead radiometric dating of the zircons from these localities tell us several things. Firstly, the Chinese samples show that extinction events started at around 251.94 million years ago and they lasted for a few tens of thousands of years. Secondly, the eruption of the Siberian Traps commenced with a period of explosive volcanism 255.21 million years ago. Thirdly, eruption of basalt lava flows and associated intrusive activity in the underlying rocks commenced 252.24 million years ago and was sustained until just after the mass extinction. Thanks to zircons, it is clear that by the point of the onset of the mass extinction, a tremendous quantity of lava, ashes and volatiles had been erupted.

We've already seen how, in the search for early life, carbon isotope studies have proved useful, since living things preferentially take in the 'light' isotope, carbon-12, rather than carbon-13. As well as giving insights into potential Archean fossils, carbon isotope ratios have their uses in deciphering mass extinctions. Generally speaking the balance maintained by the slow carbon cycle is reflected in a relatively stable carbon isotope record, but at times of mass extinctions it often shows major perturbations. This is the case with the end-Permian extinction.

A large disturbance, known as an 'excursion,' in the carbon isotope record, is found in limestones deposited at the time of the extinction, as far apart as China and Italy - and pretty much everywhere in between. The excursion shows that an enormous light carbon release occurred. One likely cause was the release of excess carbon-12 through mass dieback of vegetation. Another was the massive release of 'thermogenic' carbon-bearing gases due to the tremendous volumes of magma working their way up through that Siberian sedimentary basin and thoroughly cooking those extensive coal deposits.

A quiet day at the Siberian Traps - it was like this or worse almost every day for over 10,000 years. and the view would have been similar in all directions. Mankind has never seen volcanic activity on this scale. →

A geological smoking gun, in the form of a block of coal, one of many encased in the Siberian Traps volcanic rocks, in this case exposed in a quarry near the Angara River, Russia. Image: Scott Simper, working with an international team of Earth Scientists led by Prof. Lindy Elkins-Tanton of Arizona State University.

Finally, you can't beat a geological smoking gun and that's exactly what has been found in Siberia. Over six fieldwork seasons, an international team of geologists mounted expeditions to the district, much of which is remote, requiring access by helicopter or, along the larger rivers, by boat. In a peer-reviewed paper published in the scientific journal Geology in June 2020, the team described extensive deposits of volcanic ash, hundreds of metres in thickness and in places containing numerous fragments of burnt wood and coal. These findings clearly indicate the magmas had interacted with coal seams deep below the surface. Not only had the volcanism torched the forests of the region - volcanic eruptions often do such things, but it had also brought up colossal quantities of coal debris from the depths: the place must have been like Mordor on steroids at times. Estimates of the carbon content of the organic-rich rocks in the sedimentary basin, intruded by the rising magma, are in the order of 10 trillion to 100 trillion tons - either way these are enormous numbers. So is 36 billion tons - our annual pre-pandemic carbon dioxide emissions.

Through multiple geological techniques, it has been shown with little doubt that the greatest extinction in the fossil record was caused primarily by pollution. We may not be able to travel back in time to see what actually happened at the end of the Permian, but we continue to gather hard geological evidence. Our advancing scientific techniques mean that the rocks are giving up more and more information about past conditions, with each passing decade.

In the case of the end-Permian, the rocks point to Earth's biosphere being massively clobbered from all directions, repeatedly: lethally rapid global warming, acidic rainfall, dead plant communities, food chain disruption and starvation, massive soil erosion, toxic, acidified seas, all contributing to making conditions inhospitable to life itself. I don't think I need go into any more gory detail than that - the fossil record says enough on its own.

The take-home point from this look at Large Igneous Provinces is that the slow carbon cycle may, at widely spaced intervals, be severely disturbed by geological events. Yet here we are, busily creating such a disturbance voluntarily. Events surrounding the Siberian Traps eruptions tell us why - in no uncertain terms - this is a really bad idea.

Snowball Earth - what happened?

In the late Proterozoic, during the appropriately-named Cryogenian Period (720-635 million years ago), there was a remarkable series of global freeze-ups. Again, we're stepping a little out of the missing years in order to take a look, so extreme and fascinating were these events. Known collectively as the Snowball Earth glaciations, they featured ice sheets apparently extending all the way to, or near to, the Equator. In the first glaciation, the Sturtian, the planet seems to have stayed frozen-over for almost 60 million years, whilst in the second, the Marinoan, the cold lasted for up to 15 million years. Even ordinary ice ages are not at all common along the geological timeline: that infrequency in turn suggests their triggering must involve unusual circumstances.

Geologists think the chain of events that led to the onset of such massive glaciations began around 850 million years ago, when a great supercontinent, Rodinia, was situated over Earth's equatorial regions. When Rodinia began to rift apart in response to plate tectonic forces, there were prolonged and voluminous outpourings of basalt lava. The volcanism may not have done the catastrophic damage of the Siberian Traps, but what it did do was make available vast areas of reactive basalt - in a tropical climate. If you wanted to create the optimum scenario for large-scale weathering and prolonged CO_2 drawdown over many millions of years, that would be a way to go about it. The weathering was sufficient in its intensity and duration to cause a major fall in atmospheric carbon dioxide levels, the resultant cooling being so dramatic that it rather ran away with itself.

To get a near runaway global cooling scenario of this kind, climate feedbacks are also essential. Most importantly, the ice sheets themselves reflect incoming sunlight straight back out to space. The fraction of sunlight reflected by any surface is expressed as its 'albedo', which has a scale from one to zero. Dark surfaces, absorbing all incoming sunlight, have an albedo of zero, which is why wearing black clothes outside on a warm, sunny summer's day will make you feel even hotter. In contrast, surfaces that reflect all incoming sunlight have an albedo of one.

Let's apply this to a planet that is gradually freezing over. Open ocean water has an albedo of just 0.06, so it mostly absorbs the incoming solar radiation. But freeze that sea water into ice and the albedo moves up to around 0.5. Drop a fresh fall of snow onto the ice, making it a pristine white, and the albedo will shift up again, so that the surface will now reflect some 90% of the sunlight back out to space. So once extensive ice sheets have formed, there is a big drop in the energy received by Earth's surface. That, in turn, leads to further cooling. More cooling lets even more extensive ice sheets form, thereby reflecting still more incoming sunshine – and so on: the effect just builds up and up.

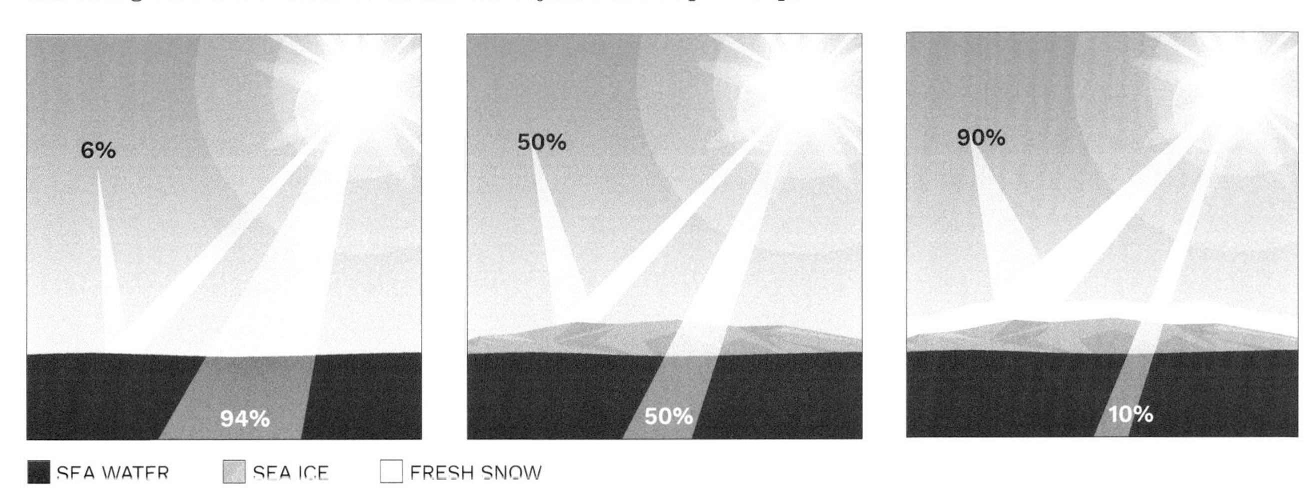

It was in the 1960s when scientists calculated that if ice sheets were ever to cover half of Earth, the albedo feedback effect would completely run away with itself, leading to ice spreading all the way to the equator. They did not, however, believe that such a runaway feedback had ever occurred. At the same time, a geologist based at Cambridge had an ever-expanding dossier of evidence for glaciations, of near-global extent, in the late Proterozoic. As sometimes happens across the scientific disciplines, the two parties – the one making the observations and the ones having the explanation for it – were for a time unaware of each other's work.

The geological evidence for glaciation initially took the form of highly distinctive sediments, transported and deposited by glacial ice. Such sediments are, with practice, fairly easy to recognise. Ice differs from wind or water as an agent of sedimentation in that it flows very slowly but powerfully and only deposits stuff when it melts. Wind and water tend to produce sediments that are said to be well-sorted, meaning that as the flow speeds up and slows down through time, layers of sediment are deposited which exhibit a larger or smaller grain-size respectively, reflecting those changes in environmental energy. Not so glacial ice. Sediment grains of all sizes, from the finest mud up to house-sized boulders, get trapped in the ice and transported, sometimes over great distances. Wherever the ice melts, the whole lot simply gets dumped in an unsorted muddle. The dumping may take place on land or out to sea. Such muddled up deposits are present, in rocks of Cryogenian age, on almost all continents.

Although the Snowball Earth events repeatedly turned the planet into an iceball, the system restored itself on each occasion, thanks to the slow carbon cycle and plate tectonics. Where would we be without them? Just as weathering accelerates in warm conditions, it slows in a cold climate. During Snowball Earth events, when many places on Earth would have looked like the deep interior of modern day Antarctica, weathering must have pretty much ground to a halt. As a consequence, the removal part of the slow carbon cycle was effectively switched off.

Despite some planetary processes being interrupted, for plate tectonics and its inevitable by product, volcanism, it would have been business-as-usual. Towards the end of Snowball Earth glaciations, we thus have an unfolding scenario where atmospheric carbon dioxide continues to be replenished towards pre-glacial levels through subduction and associated volcanic activity, whereas its removal by weathering remains insignificant. These twin factors quite literally came to Earth's rescue, reviving the planet, through the enhancement of the greenhouse effect, back into a healthy state. Plate tectonics does not occur at the same steady rate through time: at some times a lot more plate movement, subduction and volcanism are going on than at others. This point may help explain why the Sturtian and Marinoan glaciations had different durations.

It seems likely that the end of each Snowball Earth event was brought about by carbon dioxide levels building to a critical threshold, beyond which the cooling effect of the reflective ice was overwhelmed. In each case, it seems that the end came about rapidly, in geological terms, with a sudden climatic flip back to much warmer conditions. Massive scale melting of the ice sheets led to rising seas quickly flooding back in over the continental shelves, with widespread deposition of limestones and dolomites, the latter being carbonate sediments that also contain magnesium. Such rocks, known as cap carbonates, directly overlie the distinctive glacial deposits in many places.

A rock outcrop in the Kaokoveld, Namibia. Paul Hoffman, a key "Snowball Earth" researcher, is pointing to the tillite-cap carbonate boundary. Photo: M.J. Hambrey

Climatic flips such as those causing the Snowball Earth glaciations must have presented a serious challenge to the life that was present on Earth at the time. The original modelling of such freeze-ups suggested that life should not have survived them. But that work was done decades ago. More recently, discoveries of thriving communities of micro-organisms, in the cold, dark depths of the oceans, around boiling hot underwater hydrothermal vents and within sea ice, have changed that thinking. Life was here to stay and in the final period of the Proterozoic, the Ediacaran (635-541 million years ago), there was one of the occasional spurts of evolution we see, marked by a faunal diversification in the fossil record.

Late Ediacaran fossils are known from a number of places worldwide and recently, researchers have found particularly well-preserved examples containing cholesterol-like molecules, known from most modern animals but scarce in other primitive life-forms. Furthermore, we have evidence in the form of burrows in seabed sediments from this time, that there were things, possibly forms of marine worms, that could crawl around and dig themselves holes to dwell in. We may not have found their body-fossils just yet, but we have their homes to study.

Earth's atmosphere at this time, although oxygenated, was yet to become oxygen-rich like that of the present day. Estimates vary, but it seems likely that the air contained between a tenth and two fifths of the modern value of 21%. A well-oxygenated deep ocean might indeed sound like a tall order under such conditions, but there were still those photosynthetic algal mats, encrusting the shallow seabeds. It is thought that, where present in high concentrations, these 'mat-grounds' constituted localised biological oases, areas in which both the water and the uppermost sediments of the seafloor were well-oxygenated.

Thanks to that oxygenation, aerobic ecosystems could develop and flourish. Life itself thus created the right conditions for other life-forms to develop. And so, from still-unknown precursors, there began the evolution of the animal kingdom – our direct ancestors.

Dickinsonia Costata, an iconic fossil from Ediacaran sedimentary rocks of the Flinders Ranges, South Australia. There is increasing evidence these were animals that grazed on algal mats.

Image credit: Dr Alex Liu, University of Cambridge

WALES THE MISSING YEARS

Part nine: The Missing Years come to an end

Here, close to the closure of the Proterozoic Eon, the story of Wales itself, as recorded in great detail in its rocks, begins. Deep in Earth's Southern Hemisphere, off the northern margin of the great South Pole-straddling continent of Gondwana, a destructive plate-boundary came into being. Oceanic lithosphere began to be subducted in a southerly direction beneath northern Gondwana. Magma formed in vast quantities, rising from the mantle up through the overriding lithospheric plate. On the continental shelf, extensive arcs of volcanic islands began to emerge, growing in size as new crust was created in abundance by both eruptive and intrusive activity. Through tectonic movements, these new bits of crust eventually became amalgamated together, like a geological collage, to form the elongated microcontinent that geologists have named Avalonia. Wales, or proto-Wales if you like, was situated close to its middle.

The remains of these late Proterozoic volcanic arcs can be seen here and there around Wales, peeping up through the much more widespread cover of younger rocks. By studying these outcrops, attacking them with our arsenal of modern geological tools, we have already learned an awful lot about them. There is, of course, plenty more to discover, since science is a continuous process. But we know that late Proterozoic rocks form the geological 'basement' to much of the country – if you had an unlimited drilling budget, you could put a deep borehole down in most parts of Wales and eventually hit the stuff.

Unfortunately, unlimited drilling budgets are about as easy to find as unicorns, especially in these austere times. So there are still some big questions to answer. We know the late Proterozoic volcanic arcs were continental in nature from their geochemistry. That means they developed on even older continental crust. But that older crust is not exposed anywhere in southern Britain. There are indications, based on geochemistry, that much older rocks, perhaps 1,600 million years old or more, may underlie southern Britain at great depth. It's just that we probably won't be holding samples of them in our hands, any time soon.

What subsequently happened in Wales further along the geological timeline, about which we know an awful lot more, will be told elsewhere. But by the time Wales became a tangible entity, Planet Earth was already a relatively safe place, most of the time. It had plate tectonics in full swing, providing countless varied habitats for life to colonise. A generally stable greenhouse effect, itself regulated by plate tectonics and the slow carbon cycle, mostly prevented the place from being too cold or too hot. Earth had an increasingly-oxygenated atmosphere, a protective ozone layer and a well-developed magnetic field keeping solar winds and cosmic rays at bay. There was plenty of wholesome water, a range of habitable climates and widespread ecosystems thrived, at least in the seas. The land was yet to be colonised. Soils were yet to form: there was only bare rock and rubble. But colonisation was now possible and would happen in due course.

Such things, making Earth "just right" for advanced life, were all products of the interlinked processes that had dominated the preceding 3,867 million years on Earth, a time about which we have had to look elsewhere for our evidence. But that was now about to change. Finally, the missing years had come to an end.

Epilogue

This account of the formation and evolution of our only home, a planet capable of supporting advanced life, was researched, written, illustrated and edited during the pandemic-enforced lockdowns in 2020-21. For much of the first lockdown period, weather conditions were excellent. After a day's writing was done, I would often sit in the field behind my home, a grassy hillside overlooking the Dyfi Estuary, where I watched and listened to the world around me.

Society, as we knew it, had suddenly been switched off: what I recall above all else was the absence of man-made noise. You could hear the gentle calls of waders on the intertidal mudflats, nearly a kilometre away. On those still evenings, birdsong was almost unpunctuated, where normally, even here in rural Wales, there is the faint rumble of vehicles on distant A-roads, interspersed with the occasional screech of a passing high-powered motorbike. It was a rare glimpse back into another world, a world where peace and quiet reigned once again. Gone was the headlong rush in pursuit of stuff we do not need but have been trained to want, sold to us by raising consumerism to a quasi-religious status.

I guess the purveyors of stuff, the corporations who trundle it out by the kiloton, branded and over-packaged, bombarding us every waking hour with clever incitements to buy it, would much prefer that we simply returned to carrying on as usual. Buying, using for a while then throwing away. There are two problems with this.

Firstly, where is "away"? It is not in the heart of a star, where matter is instantly vapourised. In consumerist parlance, "away" is simply somewhere we cannot see, so we do not have to witness an item's slow decay, disintegration and ensuing release of pollutants. Secondly, and more importantly, there's another blindingly obvious problem that was there before the pandemic and that remains the case to this day. To carry on as usual with our Neoliberal consumer-capitalist system requires the Infinite Growth Paradigm to be correct. It is not: it's a contradiction in terms.

It was back in the 1970s that I discovered first-hand that the Infinite Growth Paradigm was a fantasy. As a small boy with a deep fascination for all things natural, I used to do all sorts of experiments, from raising butterflies or moths from caterpillars I had found, to pottering about with my chemistry set. My mother used to bottle surplus fruit for the winter, a widespread household practice back then. Very occasionally, a Kilner jar would not seal as it should and mould would set in. Discovering one such jar in the pantry, I purloined it, for the purpose of scientific observation.

Stood there on my bedroom windowsill, the jar became home to the most magnificent culture, a carpet of pinhead-topped delicate threads standing an inch or more in height above the surface of the stewed apple that was its food supply. The mould worked its way through that food over several months. It seemed unstoppable. But a time came when the food source was used up and, as quickly as the culture had appeared, it dwindled and then vanished without trace.

That mould was living the Infinite Growth Paradigm like there was no tomorrow. It was the Ultimate Consumer: taking the resource and processing it, without a care – and one cannot blame it as such, given that as far as we know, moulds do not think. That should be where we differ. But seemingly, it is not.

Listen to the endless rounds of interviews on current affairs programmes on the radio or TV and I bet you that it won't be long before some politician or other utters a sentence with 'growth', 'hardworking' and 'consumers' juxtaposed: these are the sacred keywords of today's world. None of those politicians, with a few honourable exceptions, will ever admit that the very same parameters that applied to my jar of mould likewise apply to us, for in a finite world, you cannot have infinite consumption-based growth.

Some things are incredibly abundant on Earth. Take quartz, for example. It makes up over ten percent of Earth's crust: many sandstones contain 90% or more quartz. But that's a fat lot of use to us. We cannot breathe it, eat it, drink it, weave it or set fire to it - air, food, drink, clothing and heat are five of the six absolute baseline necessities for our continued existence. Yes, sandstone can be used to build the sixth - shelter - but like the glass bottom of that mould's jar, it is otherwise as good as inert. The same point applies to many of the natural substances found on the planet.

Resources available and useful to Mankind fall into three categories. Non-renewable ones are things like the fossil fuels we extract, refine and burn, or the phosphate we mine and use for fertiliser: these take geological time spans to replenish. Renewable resources comprise tides, wind, sunlight, water power and so on, but all require the involvement of non-renewables in their harvesting. Finally there are potentially renewable resources: these are a diverse bunch of things from soils to fish stocks and everything in between. These are resources that, if managed with long term sustainability borne in mind, can renew themselves on a constant basis year-on-year. But if taken out in vast excess, through mismanagement or purely for short-term financial gain, they will dwindle away. And that's the problem: in our headlong race for infinite growth, we are going through the non- and potentially -renewable resources as if their finite nature is inconsequential. In our Consumerist world, we are acting out a similar role to that mould, albeit in a more evolved manner, as we attack Earth's stable climate, its water resources, its oceans and biodiversity on multiple fronts, seemingly without a care.

Lockdown temporarily slashed our carbon and other emissions: the air cleared of noxious pollutants, even in the cities. It was a far more drastic form of change, that most of us willingly agreed to go along with, compared to the longer-term adjustments we need to make in order to stabilise the climate and permanently deal with the pollution that claims so many lives.

Can we ultimately make those changes? If you have read this book and not simply skipped to this section, you will by now have realised it is little short of miraculous that we are here at all: everything we take for granted all around us is the product of fortuitous, interactive geological, chemical, physical and biological processes that took place way back in deep time. Had just one or two of those processes not happened, Earth would not have evolved into a place habitable by advanced and diverse multicellular life.

Such processes are therefore of incalculable preciousness to us. Their continued functionality ought to be the central motivation behind government policymaking the world over. Instead, priority number one is growth, growth, growth, regardless of the consequences to the natural world. As long as the numbers look good, nothing else really matters. Why do we place numbers on a pedestal above all else, be they pounds, dollars or yuan, when without a functioning environment they lose all meaning?

Mass extinctions, demonstrated clearly in the fossil record, show us how Earth's environments have malfunctioned on occasions in the past, sometimes on a frightening scale. There have been sudden and dramatic changes in climate. Polar ice caps have come and gone: landmasses have been inundated and have emerged from the waters. Volcanic eruptions, on a scale never witnessed by Mankind, have polluted and exterminated. Everyone has heard of Earth's periodic and devastating encounters with asteroids. A recurring theme, however, is that in each such instance, biodiversity resurfaced. It had to take its time on occasion because the environment was so badly damaged, but in every case a healthy planet was restored over millions of years. Replacement ecosystems would be without some or most of the old life forms but populated with those that evolved from the survivors. The fact that we exist is almost certainly because the mammals were able to diversify to a far, far greater extent following the K-T boundary extinction: they no longer had all those dinosaurs to compete with and hide from.

Make no mistake: we've messed up a lot of things, but given enough time, Earth will again recover from the damage we have done to it in our relentless pursuit of money, growth and power. It is a remarkably resilient planet – over geological time spans. The minuscule blip on the timeline, represented by the industrialised world of the past few centuries, will leave its mark in the geological record through its fossils, its trace element geochemistry, its stable isotope ratios and all the other indicators that modern geologists examine. It may well be that the rocks will show that a mass extinction occurred about now, for biodiversity is in many places coming under massive strain. But it doesn't have to end that way.

As I looked about me on those calm lockdown evenings above the Dyfi Estuary, I thought of the things we could accomplish once enough people grasped, with the help of hard scientific evidence, the absolute preciousness of our very existence and how fortuitously it came about. Perhaps at that point the task to win over hearts and minds, here in Wales and elsewhere, would become so much easier, to the point where the majority of people agreed that the health of the environment equates to the health and well-being of all. And where one country starts, and the results look good, others tend to follow.

I sat there, on a favourite slab of slate, and thought, "what if?"

Acknowledgements

This book was initially conceived in late 2014, then as part of a much bigger project - a detailed analysis of our geological origins here in Wales. That is an enormous topic of Tolkeinesque proportions and after putting together a framework of key events, the decision eventually came to break one potentially huge volume into a number of shorter ones. Each volume could then be a narrative of how a place or district in Wales came to be and where that prehistory fitted into the overall scheme of things.

My first volume in the series, The Making of Ynyslas, duly arrived in late summer 2019 and I set about marketing the book, learning on the hoof, but building up steady sales - until the pandemic came along. Lockdown ensued and at a stroke I lost most of the network of retail outlets I had established - a picture familiar to all too many in this sector. A swift decision was taken: to tackle this part of the story, by far the most challenging of all. That is because of the amount of geological time involved, the great complexity of the older rocks on Earth and the sheer volume of up-to-date research that would need to be understood and then processed into a digestible format. In those respects, this was an ideal lockdown project.

Preparation of this book involved talking to a lot of scientists, too numerous to mention, but to whom I owe an immense debt of gratitude for helpful discussions, clarifications, copies of their research papers and images. For their constructive reviews of early drafts of the text and their help with selection of graphics, major appreciation goes to Doug Bostrom, John Garrett, Dan Bailey and other members of the Skeptical Science team (https://skepticalscience.com).

Thanks to my fantastic designer and illustrator, Sara Holloway, for her diligence, patience and timely work, and also to my editor, Fiona Mason, for her guidance and indeed inspiration into getting this project off the ground. The support of Arts Council Wales, through the Welsh Government, has kept very many creative freelancers going through what would otherwise have been incredibly difficult times in 2020-2021 and is gratefully acknowledged. I hope this production justifies their faith in me.

John S Mason, MPhil, is a geologist by training with a long-held love of wildlife and a deep interest in weather and climate. On climate change, he writes for the award-winning website, Skeptical Science, with a particular focus upon research into climates of the past. His first book in this series, The Making of Ynyslas, was published in 2019. He is also the author of the successful Shore Fishing - a Guide to Cardigan Bay (Coch y Bonddu Books, Machynlleth, 2013), Introducing Mineralogy (Dunedin Academic Press, 2016) and numerous peer-reviewed contributions to British mineralogy, including co-authorship of the major volume, Mineralization in England and Wales (published in 2010 by the Joint Nature Conservation Committee).

www.geologywales.co.uk

Further reading

This account was based on a mountain of peer-reviewed literature. The work involved tracking down and reading dozens of papers and then checking responses to them, in order to ensure the science was as up to date as possible. Corresponding authors were frequently contacted for further discussion and to ask for digital reprints or images.

Important papers tend to get highlighted in parts of the news media and with a link to follow to the paper itself, the corresponding author may readily be found. However it's always worth entering the paper's title into Google Scholar (https://scholar.google.co.uk/) as many institutions make such papers available via that platform so time may be saved.

Papers can be highly technical simply because they often deal with investigations into areas of great complexity. This book was always intended as a bridge between that technical world and our everyday one. However, if any reader is inspired enough to pursue these topics further (and I hope they are) then there are many good entry-level and more advanced undergraduate textbooks out there. One highly effective way to locate and find out about such volumes is to search for and check out university bookshop websites as they will often list recommended textbooks for courses in the Earth Sciences.

Books sold through such outlets are likely to be reliable sources of scientific information: the reviews will soon weed out those that are not. Unfortunately, the same cannot be said for online video and social media channels, since the claims made at such sites may be baseless, evidence-free assertions of opinion. We have seen an eruption of such material, for instance, during the SARS-CoV2 pandemic. This is perhaps the biggest problem with the internet: anyone can say anything and a lot of such material then gets uncritically repeated, spreading through the network - just like a virus. Indeed, a website I work for, Skeptical Science (https://skepticalscience.com) - run by an international all-volunteer team - was founded in 2007 with the specific goal of rebutting widely-circulated misinformation on the subject of climate science - and has been busily engaged on that important task ever since.